Calochilus richae   **BALD-TIP BEARD-ORCHID**

This orchid was 'lost' for forty years. Discovered and
described in 1928 it was not seen again until 1968. Plants
are known from only one extremely localised area in
Victoria (first published colour photograph)

Cryptostylis leptochila and pollinating wasp. *(Lissopimpla
semipunctata)*.

AUSTRALIAN NATIVE ORCHIDS

IN COLOUR

AUSTRALIAN
NATIVE ORCHIDS
IN COLOUR

Leo Cady and E. R. Rotherham

REED

First published 1970
Reprinted 1977, 1978, 1981

A. H. & A. W. REED PTY LTD
53 Myoora Road, Terrey Hills, Sydney
68-74 Kingsford-Smith Street, Wellington 3
also at
Auckland and Christchurch

National Library of Australia
Cataloguing in Publication data
Cady, Leo.
  Australian native orchids in colour.
  (Reed colourbook series).
  Index.
  First published, Sydney: Reed, 1970.
  ISBN 0 589 07011 8.

  1. Orchids — Australia. I. Rotherham, Edward R.,
  joint author. II. Title. (Series).
584 150994

Printed and bound by Kyodo Shing Loong, Singapore.

CONTENTS

INTRODUCTION

THE ORCHID FAMILY, known to botanists as the Orchidaceae, exceeds in its number of species all other families of flowering plants. The number varies greatly with different authorities, and the total ranges from 15,000 to 30,000 species, grouped into several hundred genera. Many of these species have been created by the enthusiast orchid-grower.

The diversity in form of orchid flowers and their cosmopolitan distribution gives them a popularity among botanists and horticulturalists, professional and amateur. This popularity is also due to the structure and beauty of the orchid flower and its ornamentation, which is matched by few other flowering plants.

There are more than 600 Australian native orchids, which are classified into more than seventy genera or related groups. The different distributional trends in the Australian orchid population is explained in the following way. Many of our Australian genera—*Pterostylis* (Greenhoods), *Acianthus* (Pixie Caps), *Caladenia* (Spider Orchids), *Prasophyllum* (Leek Orchids), and *Microtis* (Onion Orchids), to name the major genera—had their origin in the southern States of Australia, or, as suggested by Hatch,[1] in the legendary paleozoic southern continent, from which they entered Australia via Tasmania and spread north, east, and west. These have expanded geographically only very slightly in comparison to the much greater influence on the Australian orchid flora shown by the southern Asian migration.

Two-thirds of our orchid genera had their origin in the southern Asian area. Some genera with numerous exotic species have only one species in Australia, while some of our larger genera have a very scant representation in the neighbouring islands.

It is probable that most of New Zealand's orchid genera and many of New Caledonia's originated from the wind-borne seed of Australian orchids. Once established, a particular genus would then evolve to suit the local habitats.

The spread of the typical Australian terrestrial orchids falters as it reaches the tropics, where the southern Asian types take over. The great majority of our terrestrial orchids occur in the temperate zone. Many of the Australian orchids are no longer available for study as the expansion of our settlements, combined with indiscriminate and often illegal collecting, and an inability of the plants to adapt to a changing habitat, have led to the extinction of many of the rarer species.

Australian native orchids are protected plants in all States, which means that they must NOT be picked or collected from any National Park, Forest Reserve or any other Crown Land, without an official permit; nor can they be collected from private land without the owner of that land's consent.

Disregard of these restrictions can bring heavy fines.

The name orchid comes from the Latin *orchis*, which in turn has its origin from the Greek *OPXU* which refers to the testis-like shape of the tuber in some of the terrestrial forms of the plant.

Orchids have a very large and varied distribution, being found in all areas of the world apart from the polar and desert regions. Although a number of Australian species occur in semi- or very arid areas, the greatest density of species and genera occur in areas of tropical rain-forest where they reach their maximum size, diversity of form, and abundance.

Orchids are perennial plants, dividing easily into two major groups.

(1) Terrestrial orchids, which have tubers (small round fleshy underground parts of the root system) or rhizomes (an elongate fleshy root stock). They are normally found growing in sandy, loam or clay soils, or in the deep humus of the forest floor. Some genera are found on the trunks of tree-ferns or in the forks of trees, and occasionally on mossy rocks.

(2) Epiphytic orchids, are generally supported on the trunk and limbs of trees or on moss and humus-covered rocks, being more abundant in the tropical rain-forests, gradually diminishing in numbers until we find the southernmost record of any species in Stewart Island, off New Zealand's South Island. It is the epiphyte that receives the highest acclaim from the horticulturalist. For this reason the South American *Cattleya* and the southern Asian *Phalaenopis* have paid the very high price of extinction in their native habitat. Epiphytic orchids are not parasites; they use the tree or rock only as a support in order to get higher in the forest for stronger light. The crevices and bark fissures are used as channels to guide the rain and humus to the main plant.

The plant has adapted itself to this condition by developing very unusual root mechanisms. The outer layer of cells of the root is thick and absorbent and can absorb water and its mineral contents very rapidly. The root is tipped by a root cap with a bright green or whitish-green end. This covering is the actual growth point of the root. Orchids anchor themselves to the tree on which they grow by the use of special cells that form on the underside of the root. For this reason epiphytes are usually found only on trees with persistent bark; in Australia, ironbarked Eucalypts, *Melaleucas* etc. Some plants also have adapted the leaves as a means of storing food and water; the leaves in these cases are thick and fleshy. Other orchids use the stem as a storage area and we find orchids with large bulbous bases or pseudo-bulbs.

The Australian epiphytes are found at their best north of the tropic of Capricorn, but extend southwards to Tasmania where two species, *Dendrobium striolatum* and *Sarcochilus australis* occur. Some epiphytes grow mainly on rocks; these are Lithophytes ('rock lovers').

Because of the dependance of the epiphyte on moisture in the air, their main occurrence in Australia is confined to the eastern coastal ranges and areas of relatively high rainfall and high humidity. In South Australia and Western Australia the occurrence of the epiphyte is rare. *Cymbidium canaliculatum* appears to be the most xerophytic (growing in arid conditions), being found in the drier inland areas of New South Wales, Queensland, Northern Territory, North-eastern Western Australia and once recorded from inland South Australia.

Terrestrial orchids in Australia are divided into two main groups.

(1) Green-leaved herbs with alternate leaves, either stemclasping (cauline), or radiating at the base of the stem (rosulate). In some cases the flowering stem is separate to the main rosette but connected to it underground, as in *Pterostylis parviflora*, or at times the leaves either precede or postcede the flowers.

(2) Semi- or Holo-saprophytic (saprophyte means an organism using non-living organic matter for nutriment). This group contains some of the most unusual orchids in the world, *Rhizanthella gardneri* and *Cryptanthemis slateri*, see page 82.

In some genera in which the majority of the species are typical green-leaved plants, one or more

7

species may occur as saprophytes. In general saprophytic plants lack the green chlorophyll cells that manufacture the plant's sugars and starches. To overcome this, the group depend wholly on a symbiotic relationship with a soil fungus. At least two types of fungus have been recorded. The *Hymenomycete* type, with clamp-like connections to the root, and the *Rhizoctonia* type, in which the fungal hyphae (the underground growing body of the fungus) intrude into the cells through a break in the outside of the root, or through passage cells in the root's outside layer of cells. On entering the root the fungus coils itself into a tight spiral of hyphae within the roots cells.

Within the root structure are three different types of cells. One is called the fungal host cell, where the fungus invades the plant; cell two is the digesting cell and the third a storage cell layer, stores the starches acquired from the digestion of the fungal hyphae. When the fungus enters the root or rhizome of the plant, it infects the first cell layer and grows rapidly. The plant then intervenes and digests the fungus within the second series of cells, storing the resultant products in the third layer of cells for the plants' later use.

This symbiosis is not confined only to saprophytes, but occurs to a lesser degree in almost all of the orchid genera, including the epiphytes, and it has been shown that in many cases each species or related species of plant uses a particular type of fungus.

Fungi also play a major part in seed germination in nature, but it has been proved that the fungus is unnecessary if the seeds are grown by the aseptic agar jelly method used by most commercial orchid growers.

The orchid family is classified as the highest evolved form of plant life. The reason for this is the orchids' floral structure and the plants' specialised methods of attracting insects for the pollination of its blooms.

The contents of this volume have been arranged to give a broad coverage of the majority of the common Australian orchids. While many of the plates depict orchids occurring in several States, there are also several illustrations of rare and very rare, often extremely localised plants.

Many people have an aversion to the use of the scientific or botanical names bestowed on a plant, these names are of utmost importance for they are universally accepted and recognised and are changed only after considerable research. The common or popular names of a plant often differ from one district to another and occasionally become applied to entirely different groups of plants. For instance, Australian Spider Orchids are a group of plants in the genus *Caladenia*, but the same common name in New Zealand refers to a totally different genus, *Corybas*. The representatives of this genus in Australia do not have the long sepals and petals of the New Zealand plants.

To English orchid-hunters the Early and Late Spider orchids refer to species of the genus *Ophrys*. The flowers of the Australian *Dendrobium tetragonum* are also occasionally referred to as Spider Orchids. The confusion that can arise from a 'common' name is thus considerable and the desirability of using the scientific name of the plant becomes obvious.

The scientific or botanical name of a plant is made up of two parts e.g. *Pterostylis acuminata*, and is usually set in italics. The name is generally derived from Latin or Greek words. The first or generic name is always written with a capital letter and refers to all those plants which share certain characteristics, e.g. *Pterostylis*. The second name is an adjective, and must agree in gender and number with the generic name. It is therefore written with a small letter and is generally descriptive, e.g. *acuminata*, meaning 'pointed'. Occasionally distinct differences occur within a species and the species is divided into varieties or forms, e.g. *Pterostylis acuminata var. ingens*. The botanist further adds after the name the initials (or name if living) of the person who first described the plant, and if after its original description it has been renamed or reclassified, the initials or name of the author of this reclassification. Thus correctly written the botanical name of the orchid used above should be recorded as *Pterostylis acuminata* R. Br. *var. ingens* H. M. R. Rupp. This records that the genus and species were first described by Robert Brown and that the variety *ingens* was established by H. M. R. Rupp. As a space economy only, the authorship of the species has not been recorded in this volume. Reference to the works listed in the bibliography (page 111) will supply this information.

The flowering periods listed are those generally applicable—orchids usually flower with an amazing regularity, but a plant occurring in Queensland may flower ahead of plants of the same species in Victoria and Tasmania. Similarly species in Western Australia may bloom before the same species flower in the eastern States. Altitude too affects the flowering period. In the alpine areas some early spring and even autumn flowering species of lowland areas are in bloom in the summer months.

Colour, height and size can often vary considerably within a species and should not be the sole method of evaluation. Adverse weather conditions can produce stunted plants—while a good season may yield robust plants.

Many methods and devices are employed by orchids to lure insects to the flower; these are discussed where known in the appropriate sections relating to genera throughout the text.

Natural hybrids do occur at both generic and specific levels; some of the commoner ones are *Dendrobium* x *gracillimum* (*D. speciosum* x *D. gracilicaule*); *D.* x *delicatum* (*D. speciosum* x *D. kingianum*); *D.* x *suffusum* (*D. kingianum* x *D. gracilicaule*); *Diuris sulphurea* x *D. maculata*); *Caladenia patersonia* x *C. dilatata*; *Pterostylis* x *toveyana* (*P. alata* x *P. concinna*). At the generic level we have *Glossodia minor* x *Caladenia caerulea* and it has been suggested that the very rare *Caladenia tutelata* is the union of *Glossodia major* x *Caladenia deformis*.

The only orchid exploited commercially for reasons other than its blooms is *Vanilla planiflora*. Its use for vanilla essence and flavouring was known to the Aztec Indians of Mexico before the arrival of the Spanish in the late 1490s. The essence is extracted from the bean-like seed pod. In commercial production the plants are hand-pollinated and while still grown in tropical countries the synthetic manufacture of vanilla essence has made serious inroads into this industry.

9

THE ORCHID FLOWER

THE FLOWERS OF ORCHIDS are separated from their nearest relation the lilies (Plate 1) in the following way. The lily, as with the orchid, has six floral appendages, three sepals outside three petals. In the lily all six segments are generally similar in shape, but in orchids the dorsal or upper petal is modified radically and is given the special name of labellum (little lip). It is often referred to as the orchid's 'tongue', and in many cases it is attractively coloured and ornamented with various types of hair and channels, or glands. During its development the flower of the orchid, in contrast to that of the lily, twists round 180 degrees so that what is actually the dorsal petal (labellum) comes to take up a ventral (or lower) position. Thus compared to the lily the orchid flower as you view it is actually upside-down.

The reproductive parts of the lily consist of six male structures called stamens, each consisting of a filament or stem portion, topped by a pollen-bearing anther. These stamens surround the pistil, or the female reproductive structure, which consists of a base or ovary which contains the embryonic seeds and an elongated projecting style, topped off by the stigma, which receives the pollen from the male structures. In the orchid flower there is only one fertile stamen, which is fused to the pistil to form the column. In most Australian genera the column is usually a slender structure, the apex containing the anther cap, beneath which are the pollen masses, usually of various shapes. The grains of pollen are held together by elastic-like filaments and the pollen masses are then called pollinia. In some genera these pollinia can contain a strap-like appendage called the caudicle, which in turn is attached by a viscid disc to a beak-like projection called the rostellum. Situated immediately below the rostellum and of various shapes is the stigma or stigmatic plate where the pollinia or some pollen grains must be placed to fertilise the ovary. The ovary is immediately below the column and is said to be inferior (below the flower). Inside this ovary are found the ovules or embryonic seeds, in vast numbers.

Fertilisation occurs after the pollen grains are placed on the sticky stigma. The grains germinate and grow, sending a tube down inside the column until it enters one of the ovules, thus effecting fertilisation. After a given period, from a few days in *Epipogum* to many months in *Cymbidium*, the ovules ripen and the capsule splits to liberate the seeds. In some genera e.g. *Corybas*, the seeding stem is greatly elongated to aid the dispersal of seed. Orchid seeds are extremely small and light and can be carried hundreds of miles by winds.

Two methods are used by orchids to induce pollination: first, the use of an insect agent, which is the commonest method; second, self-pollination. In most cases self-pollinating plants rarely open their flowers except in hot, humid conditions. The column is similar in both cases except that the pollinia in the self-pollinating species become friable in the early bud and scatters pollen onto the stigma before the flower opens, whereas the pollinia in the cross-pollinating species remain together in their groups, attached to the viscid disc for easy removal.

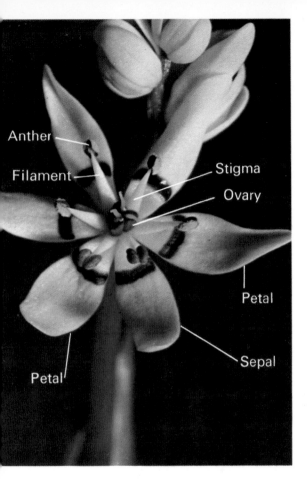

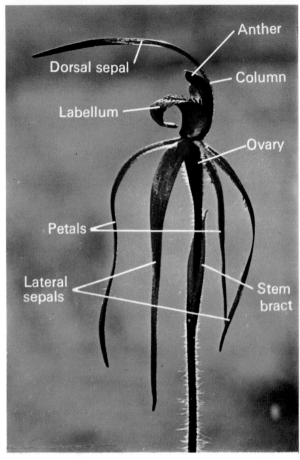

Plate 1

Anguillaria dioica   EARLY NANCY
A widespread, spring-flowering native lily.

Plate 2

Caladenia patersonii v. concolor   RED SPIDER ORCHID
Though of limited Victoria and New South Wales
occurrence, many floral features of orchids in general are
depicted in this flower.

Habenaria

THE NAME HABENARIA is derived from the Latin *habena*, meaning strap or rein, which alludes to the shape of the spur of the flower. This is the only Australian member of the large European orchid group *Basitoneae*, which has two anthers. If one is familiar with the normal Australian orchid it is indeed strange to see the column of this orchid genus with the two anthers, one either side of the column. This is a polymorphic genus of approximately 500 species, occurring chiefly in tropical areas of the world.

None of the Australian species have large or showy flowers and none of our species is at all common in the damp areas of open coastal forest, which these orchids seem to favour.

Approximately sixteen species occur in Australia, nine confined to Queensland, four to the Northern Territory, two to Queensland and the Northern Territory, and one to northern Queensland, the Northern Territory, and north-western Western Australia.

The Australian plants are chiefly terrestrial herbs with fleshy ovoid tubers. The plant is simple with basal or stem-clasping leaves and the small flowers are in a short raceme. The sepals and petals are free and variable in shape while the dorsal sepal is erect or incurved to form a hood over the column. The labellum is trilobed, entire (smooth) or toothed on the edges, extending at the base into a spur. The column is short. The two stigmas are usually separate or on elongated processes on either side of the base of the column and are often joined to the base of the labellum. Two pollinia are present.

The exotic members of this genus respond well to cultivation, and our endemic species could also be as tolerant. But owing to the rare nature of the plants this is difficult to forecast.

Monadenia and Epidendrum
NATURALISED ORCHIDS

MONADENIA IS DERIVED from the Greek *monos*, one, and *aden*, a gland. A genus of about nineteen species mostly concentrated in the Cape Province of South Africa, one species, *M. micrantha*, is found in the King George's Sound region of Western Australia.

This *Monadenia* was described erroneously in the *Australian Orchid Review* by the Rev. Rupp as *M. australiensis*. The orchid probably arrived in dust on sacking, etc. from the covers over shipped cargo unloaded at Albany. Alternatively the seed could have clung to the feet of wading birds, the distance being too great for wind-blown seed dispersal, even though the prevailing winds come from the direction of South Africa.

M. micrantha occurs in Western Australia at Albany, Young's Siding and Brussleton. It is a robust plant up to 40cm high, densely-leaved at the base and extending into a densely-flowered stem which in turn is densely covered with long bracts. The small greenish-yellow flowers are numerous on the spike. The reddish hooded dorsal sepal has a spur at its base. The flowering period is from October to November.

Another plant that has naturalised in the southern Queensland area is the American *Epidendrum o'brienianum*, known commonly as the Red Crucifix Orchid. It is a tall scrambling plant with thin reed-like stems and leathery alternate leaves. The flower head is an elongation of the reed-like stem. The flowers are produced over a long period, and many of them produce seed capsules. Pollination could be carried out by the ants that constantly visit the plant. Flower stems should not be picked as they continue to flower for months.

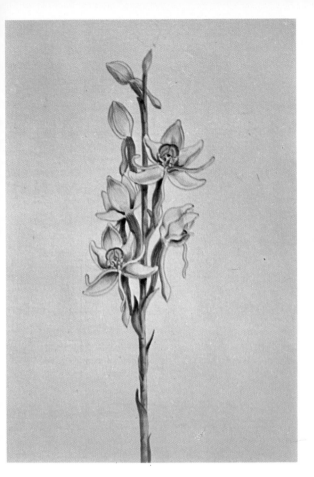

Plate 3

Habenaria ferdinandii

The Australian species of this genera are confined to the Cape York area of Queensland and near coastal areas of Northern Territory.   *Drawing: M. I. Barnes*

Plate 4

Monadenia micrantha

Of South African origin, this orchid has become well-established in south-western Western Australia. Flowers September to November.

Pterostylis GREENHOODS

THE NAME PTEROSTYLIS is derived from the Greek *petron*, meaning wing, and *stylos*, style, which refers to the wing-like extension on the column of these orchids. (Diagrams, page 20). This is a large genus of terrestrial herbs usually having small underground tubers (in some species the size almost of a 10-cent piece). The leaves are often either in a basal rosette at or near the base of the flowering stem or are narrow and stem-clasping (cauline).

Usually green, the flowers are marked with red, brown, or purplish colours—even pink in some tropical species. The unique feature of the genus is the fusion or partial fusion of the dorsal sepal and the lateral petals into a hood or galea which conceals the column. The lateral sepals may be fused for a portion of their length and their point of divergence is known as the sinus. Their total length may or may not exceed the hood. In other species the lateral sepals may be free, pointing either forwards or else deflected. The labellum is attached to the column by a strap-like appendage which gives it considerable mobility.

The genus was first described by Robert Brown in his *Prodromus Florae Novae Hollandiae*, published in 1810. In this he described the eighteen then-known species. Robert Brown spent three years, 1802–5, in Australia as a botanist and accompanied Mathew Flinders on his voyages around the continent. As a result of the above-mentioned book he is often referred to as the 'Father of Australian Botany'.

There are now seventy-eight taxa or named species of *Pterostylis*[2] which extend beyond Australia to New Zealand, New Caledonia, and to Papua.

Robert Brown divided the genus into three divisions, using chiefly the appendages and the leaves as features of separation. In 1933 The Rev. Rupp suggested an amendment to Brown's classification and gave the now accepted division of the genus.

A. *Laminatae:* In this section the labellum is strap-like, broad oblong, often narrower, usually with the edges entire but at times veed at the tip, the blade edges and appendages at times hairy. The hood curves forwards.

B. *Filiformae:* The labellum is filiform and round with long yellow hair almost to the tip, terminating in a dark knob. The Galea is erect.

Section A contains all the known species but two, and can be broken into two major and two minor groups. In the first group are: 1, plants with rosettes of leaves; and 2, plants with stem-clasping foliage. Sub-group 2 can again be divided into 2(a), those with one flower, and 2(b) those with several flowers on a stem.

The general habitat can vary greatly with the species concerned; damp semi-shaded areas on the edge of swamps and watercourses, dry arid areas or in open forest. Some, such as *P. pulchella*, occasionally occur as lithophytes on damp rocks in watercourses. Others grow close to the sea and are not affected greatly by the salt air. A very favoured habitat is under the coastal *Melaleuca* and *Leptospermum* trees, in very sandy leaf-mould. It is this leaf-mould, with the addition of a small percentage of peat-moss and charcoal, that has proved to the authors to be an excellent culture medium. *Pterostylis* are most adaptable to cultivation, with only the Rufa group and a selected few species giving any trouble.

Pollination has been described by a number of people. Cheeseman, a New Zealand botanist, was the first to publish his observations in 1873. R. Fitzgerald added further observations in 1882. These two authors are referred to by Charles

Plate 5

Pterostylis longifolia   TALL GREENHOOD
Half the galea or hood has been removed
to show how this male Mycetophilid fly
on endeavouring to escape through the
column wings of the orchid has had the
pollinia implanted on its head. Notice
the pollen granules adhering to the stigmatic plate of the orchid thus indicating
that pollination has been effected.

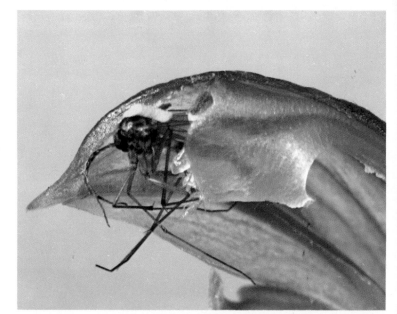

Plate 6

Pterostylis falcata   SICKLE GREENHOOD
Having escaped through the tube-like
passage formed by the labellum and
column, this insect has the orchid
pollinia implanted on its back. On
escaping from another flower the pollinia
will adhere to the stigma and other
pollinia will be stuck to the insect's back.
                    Photo: J. B. Fanning

Darwin in his famous book *The Fertilization of the Orchids*.

The first to see this occurrence in Western Australia was O. Sargent[3] who witnessed the pollination of *P. vittata* by small gnats in 1907. Edith Coleman[4] added to the knowledge of this subject in 1934.

Pollination occurs in the following way. The labellum is hinged on a strap-like appendage and is very irritable, returning quickly against the column and column wings at the slightest disturbance. Small drops of nectar at the base of the column or column foot, as it is classed, usually attract the visiting insects. The insect alights on the labellum, disturbing it so that it is thrown violently into the hood and imprisoned. The only escape route is to climb up the tube formed by the closed labellum and the column wings. Hairs are usually provided for the insect's assistance—or deterrence—as the case may be—on its journey through the 'tunnel'. At the outlet the insect must press past the pollinia with their sticky viscid disc, this segment adhering to the body of the fly, gnat, or mosquito. This same procedure is repeated with another plant, and the pollinia on the insect's back is pressed against the stigma, which is located beneath the column wings. Proceeding on its way out of the flower again the insect has further pollinia attached to its back (Plate 5 and 6).

Only three species in the whole genus appear to resort to self-pollination—one New Zealand species *P. humilus* and two very rare Australian species, *P. crypta* and *P. celans*.

After the seed capsule has formed, the seed takes approximately three months in most cases to ripen and split. Some species such as *P. cucullata* elongate the flowering stem to aid in the wind dispersal of the seeds.

Variegation is rare in orchids, but a specimen of *P. nutans* has been seen with distinct variegation on the flower and leaf rosette. This feature was not, however, repeated on the next new growth.

Possibly one of the commonest Greenhoods would be *P. nutans*, the Nodding Greenhood (Plate 8). Its distribution is wide, embracing all States of Australia and New Zealand (although the Western Australian record is very doubtful). It has a smaller New South Wales variety *hispidula*. *P. curta* is also common in the eastern States and is distinct with its blunt apple-green hood and the very distinctive twist to the labellum. Another widespread species is *P. nana*, occurring in all States except Queensland, and also in New Zealand, where it was known for many years as *P. puberula*. Generally this is a plant of rather small habit in the eastern States but in Western Australia it can reach a height of 15cm. The largest species is *P. baptistii*, a large plant which reaches 40cm in height, the hood being up to 6cm. Only two other species can approach this size—*P. falcata*, with its sickle-shaped flower of light green and the New Zealand *P. banksii* and its variety *patens*.

Of the cauline leaf group (plants with stem-clasping leaves) *P. obtusa* is widespread in the east with *P. robusta* the western and southern counterpart. *P. decurva* (Plate 7) replaces *P. obtusa* in Tasmania.

One of the most beautiful of the cauline group is the superb Greenhood *P. grandiflora* (cover) which occurs in Tasmania, Victoria, New South Wales, and Queensland, and flowers from April to August. A very distinctive plant, the petals are formed into a broad flare at its apex which is brown, contrasting with the green of the rest of the flower. The labellum is very narrow at its top half and ends in a distinct club. Another plant in this section is *P. revoluta*, a large flower 5cm high on a plant up to 28cm tall. The flower is green, marked with brown. The

Plate 7

Pterostylis decurva   SUMMER GREENHOOD

Flowering in hill and mountain country of Tasmania, Victoria and New South Wales from November to January.

Plate 8

Pterostylis nutans   NODDING GREENHOOD

Winter to spring flowering in all States except Western Australia and Northern Territory, also New Zealand.

labellum is ovate-lance-shaped, the apex acute, gracefully curving through the sinus of the lateral sepals. *P. reflexa*, a smaller, more 'square' flower; the labellum is very dark brown, narrow and lance-shaped, tapering to a long fine point.

A localised species is the Red Hood, *P. coccinea*, of New South Wales only; a very handsome but rare species found sparingly on the tablelands in open grassy places. Its flower is large, up to 4cm, greenish with bright red suffusions and bands. *P. alata* is a smaller species of slender growth up to 26cm high, resembling *P. obtusa* somewhat but the union of the lateral sepals is higher and its labellum is broader at the base, markedly constricting in the upper third. This species was one of the earliest-named Australian orchids, collected and named by the French botanist Labillardière as *Disperis alata*.

P. alata occurs in Tasmania, Victoria, South Australia, and New South Wales. In this section we find three presumed natural hybrids, *P. toveyana*, a plant originally discovered at Mentone in Victoria in 1907, in close association with *P. alata* and *P. concinna* but never in any quantity. It has taken the habit of *P. alata* to a great extent, but exhibits the bifid labellum of *P. concinna* to a small degree. *P. alveata*, now an accepted specific name, is presumed to be a union between *P. concinna* and *P. obtusa*. It closely resembles *P. obtusa* with the labellum tip variously notched in the Victorian form and bifid in the New South Wales form.

Another supposed hybrid from Section I recently published is *P. conoglossa (P. concinna* x *P. ophioglossa)* from New South Wales material, which closely resembles *P. ophioglossa*. Another hybrid between *P. revoluta* and *P. pulchella*, concerning a plant from Dripstone, New South Wales was mentioned by Rupp in 1946. The doubt occurs because the only known habitat of *P. pulchella* is very local, in a restricted area in the central coast of

New South Wales in very wet conditions unobtainable at Dripstone, a dry western New South Wales town.

Multiflowered Greenhoods, a group that R. Brown classified as his third section i.e. 'Appendage; obtuse, undivided': consists of thirteen species, two subspecies, and two varieties. In this group occurs the Rufa Greenhoods. One of the main reasons that this group is not separated today is that all of the species still have a laminate labellum and are not really much different in labellum shape from the preceding species. They do, however, join two other species *P. parviflora* (Plate 9) and *P. recurva* (Plate 10) as a naturally separate group.

A recent revision[5] of the Rufa section has given a much clearer understanding of a rather complex group. Floral segments have been studied in one taxon *(P. gibbosa* ssp. *mitchellii)*. The position and shape of the lateral sepals varied greatly in the plant studied and could add confusion if much importance were placed on them. Briefly the variations are as follows. In forward bud and freshly opened flower the lateral sepals are pendulous and straight. During the first day the tips gradually hook forwards (as in *P. hamata*); on the second day they diverge slightly, returning to the pendulous position and losing the forward hook; from the second to the third day they take a very divergent position (resembling a waxed moustache); on the third to fourth day the flowers decurve slightly, taking the sepals well back behind the ovary. This position is held until the flower begins to close. Another point noted was that the labellum, which is usually very mobile, could not be triggered for a short time after it returned to its open position after being closed.

This alteration of segment position and shapes also occurs in the genus *Prasophyllum* and has at times been the cause of new species being erected in error.

Plate 9

Pterostylis parviflora   TINY GREENHOOD

Flowering from late autumn to early spring, this multi-flowered plant occurs in all States except Western Australia and Northern Territory.

Plate 10

Pterostylis recurva   JUG GREENHOOD

Though of widespread occurrence in Western Australia, this orchid is found in no other State. August–September.

Photo: F. M. Coate

The most distinct species in the multiflowers is *P. woollsii,* a remarkable plant having lateral sepals fully 8 to 10cm long. It is found in hilly and stony country in dry conditions on the tablelands and western slopes of New South Wales. It is also found in southern Queensland and once at Rushworth in Victoria. Approaching *P. woollsii* is *P. gibbosa,* ssp. *mitchellii,* although the lateral sepals are nowhere near as long, only reaching 20cm. The labellum is oblong-ovate, shallowly dished on the upper surface, very irritable and with approximately ten marginal hairs. The flower colour is green, tinted with brown. This species is recorded only from subtropical Australia, and the remaining records from the southern States now appear to be invalid. *P. rufa* (Plate 11) is a plant that has been until recently ill-defined. It has long been collected under the name of *P. pusilla* var. *prominens* in New South Wales and Victoria. The name 'rufa', or reddish, is very appropriate to the plant. The species with its subspecies *aciculiformis* (meaning needle-pointed, referring to the lateral sepals) are recorded from all States except the Northern Territory. *P. boormanii,* a squat robust plant with heavy cilia on the sepal-margins, is a very distinct plant, olive-green with reddish tints. It occurs in New South Wales, but is rare in Victoria and South Australia. Three 'new' species were added in the review to Rufas— *P. hamata,* with robust flowers and forward-hooked lateral sepals; *P. ceriflora,* with its very short, much reflexed lateral sepals and waxy-looking flowers, only found in the Illawarra area of New South Wales, and *P. biseta,* with acute lateral sepals, finely pointed and often widely spreading, and pale green flowers.

Two species comprise the section *Filiformae—P. barbata,* an endemic Western Australian plant and *P. plumosa* (Plate 12), a widespread plant from Tasmania, Victoria, New South Wales, and New Zealand. The thread-like ciliate labellum has no parallel in the genus.

Column labellum and dorsal sepals of: 1, *Pterostylis plumosa;* 2, *Pterostylis biseta;* 3, *Pterostylis nutans.*

A. anther
AP. appendage
C. column
CW. column wings
L. labellum
LS. lateral sepals
O. ovary
S. stigma
SB. stem bract
SC. setae (labellum hairs)
UL. upper lobe of column wings

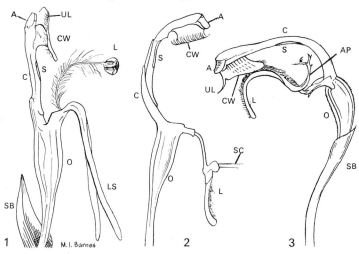

M. I. Barnes

20   1   2   3

Plate 11

*Pterostylis rufa (*syn. *P. pusilla v. prominens)* RUDDY HOOD
One of the recently revised group of 'rufa' greenhoods.
October to November flowering. All eastern and south-
ern States to southeast Queensland.

Plate 12

Pterostylis plumosa (P. barbata)   PLUMED GREENHOOD
Reclassified (1969) by L. Cady, this orchid and *P.
barbata (P. turfosa)* of Western Australia have long
threadlike hair-covered labellums. Spring flowering in
south-eastern States.

Microtis ONION ORCHIDS

THE GENUS MICROTIS, from the Greek *mikros*, meaning small and *ous* an ear, alludes to the ear-like auricles on each side of the anther.

Five species were described by R. Brown in 1810. The earliest plant of this genus was described in 1786 by Forster[6] as *Ophrys unifolia*. In 1800 Swartz[7] also named Forster's plant *Epipactis porrifolia*. In describing the new genus R. Brown included an *M. rara* which proved to be both Forster's and Swartz's plant. Reichenbach cleared up the identity of these plants in 1871.

Microtis is a small genus of fourteen widely distributed species in Australia, three of which extend to New Zealand and two to China. Of the taxa, nine are found in Western Australia.

The commonest species is *M. unifolia* (Plate 14), occurring in all Australian States, New Zealand, and north to China. It is a very variable species; the labellum calli arrangement can have the basal glands fused into one large hollow, circular gland, and the apical gland can also be lacking. The labellum of *M. parviflora*[8] is narrower at the apex and lacks the apical callus, but it has the same range as the former species. *M. oblonga* again resembles *M. unifolia*, differing only in longer petals and labellum, as long as the ovary, whereas *M. uniflora's* labellum is rarely more than half as long as the ovary. *M. truncata* is a Western Australian species similar to *M. unifolia* apart from a large hastate apical callus. *M. alba*, a beautiful Western Australian species, has the largest flower of the genus which is pure white; the apical glands are tuberculate and the apex of the lamina is bifid. *M. atrata* is the smallest species. The flowers are minute, and yellow-green. It occurs in Western Australia, South Australia, Victoria, and Tasmania.

Orthoceras HORNED ORCHID

ORTHOCERAS, FROM THE Greek *orthos*, meaning straight, and *keras*, a horn, alludes to the narrow upright lateral sepals of this orchid. It is a monotypic genus described by R. Brown in 1810. It contains only one species, *O. strictum*, with one variety *viride*. The genus has a wide distribution throughout all eastern States of Australia including Tasmania (where it is very rare), New Zealand, and New Caledonia. It flowers from November to December.

O. strictum (Plate 13) favours damp fairly swampy conditions. Unflowered plants of this genus closely resemble the genus *Diuris* but when the flower spike appears any doubt on the plant's identity is quickly removed.

The pollination of *O. strictum* was discussed by Dr Rogers[9], who discovered that it was self-pollinating. The mechanics of the action are brilliant and extremely interesting.

In the advanced bud the dorsal sepal is completely closed by the labellum. If the column is examined it is found that the stigma is oval in shape. The short diameter is placed horizontally. At the base is a funnel-shaped hollow. The pollinia lie behind the stigma base in the anther cases, and as the rostellum bends forwards the pollen masses drop out of their cases and fall on to a ridge at the back of the stigma. Pollen grains germinate there and numerous pollen tubes penetrate the rear of the stigma; at this point the rostellum collapses on to the front surface of the stigma. The funnel-shaped cavity accepts the large gland on the labellum as it returns to its bud position, pressing the sticky stigma back against the pollen masses. Thus fertilisation is effected.

Orthoceras does not take kindly to cultivation, lasting only a season or two.

Plate 13

Orthoceras strictum   HORNED ORCHID

Flowers are generally dark-green with purple-brown markings. Summer flowering in all eastern States, South Australia, and also New Zealand.

Plate 14

Microtis unifolia   ONION ORCHID

Widespread in all States except Northern Territory, also found in New Zealand. The tiny flowers need a hand lens to be fully appreciated. Spring to summer flowering.

Diuris DONKEY ORCHID

COMMONLY CALLED Donkey Orchids or Double tails, *Diuris* comes from the Greek *dis*, two, and *oura*, a tail, referring to the lateral sepals.

A large genus of more than forty species with several varieties and forms, it is almost wholly confined to Australia, with only one, *D. fryana*, endemic to Timor. New South Wales has the largest number of species, thirty-five, with six varieties. Some species are of very dubious standing. Western Australia has eight species, seven endemic and *D. longifolia*, which is found in all other States bar Queensland and the Northern Territory.

The genus was established by Sir J. E. Smith,[10] an English botanist in 1804, with the descriptions of three Australian taxa, *D. maculata*, *D. aurea*, and *D. punctata*. For many years Smith's name *punctata* was quoted as a misnomer. Certainly his plate showed a beautifully spotted plant. It was not until 1934 that P. A. Gilbert rediscovered this form in the Campbelltown area of New South Wales and the existence of Smith's plant was realised. In a recent review of *D. punctata*[11] A. W. Dockrill divides this species into six varieties and three forms.

Double tails can be found in most types of habitat, preferring open grassy patches or open forest areas, especially in New South Wales and Victoria.

The pollination of some species of the genus has been dealt with by Mrs E. Coleman[12] with the investigation of *D. pedunculata*, and *D. sulphurea*. FitzGerald included notes by A. G. Hamilton on *D. tricolor* and *D. abbreviata;* finally notes on *D. longifolia* are given by Mrs R. Ericson.[13]

In general *Diuris* use insects as pollinating agents.

Edith Coleman states that *D. pedunculata* is pollinated by a small native bee, *Halictus languinosus*. This species offers the insect the reward of sweet nectar which is found in the tissues at the end of the tube formed by the ridged sessile labellum and the column. To reach the glands the bee must force apart the segments, bringing the insect's head into contact with the viscid disc, which rests in a slot at the top of the plate-like stigma. As the bee withdraws so does the affixed disc and pollinia. In some cases, if the bee is too small, it remains a prisoner, gripped by the pressure of the labellum against the column.

D. sulphurea is pollinated in a similar manner to *D. pedunculata*, but Edith Coleman suspected from the attitude of the agents that the influencing lure might be of a similar sexual nature to the *Cryptostylis* pollination attraction.

Mrs Ericson has reported the suspected pollination of *D. longifolia* by the leaf cutter bee *Magachile remeata*.

Apart from the Western Australian species, where most species are endemic, the genus can be divided mainly by colour and by the length of the lateral sepals. The predominent colour in the eastern States is yellow or tones of yellow, often spotted or suffused with brown, red-brown or purple. Lilac and greyish-blue and white suffused with lilac are also found.

The Western Australian species can be differentiated by their leaves: in Group 1 the orchids have three or more leaves, very narrow, which can be divided into (a) leaves spirally twisted, and (b) bristle-like leaves.

Examples of 1(a) are *D. laevis*, or Nanny-goat Orchid, with pale-yellow flowers with brown blotches on the underside of the segments, and short dorsal sepal; *D. filifolia*, Cat's Face, with clear-yellow flowers and red-brown marks on the labellum only, and *D. purdiei*, which has beautiful wallflower-coloured flowers.

There is only one species in 1(b), *D. setacea*, a yellow flower with several reddish marks.

Plate 15

Diuris longifolia WALLFLOWER DIURIS or COMMON DONKEY ORCHID

Present in all States except Queensland and Northern Territory. All yellow and hybrid forms are known. Spring flowering.

Plate 16

Diuris aurea GOLDEN DIURIS

Spring flowering orchid of New South Wales and Queensland. (See *D. venosa*, back jacket.)

Orchids in Group 2 have two to three long, narrow leaves, with 2(a) the mid-lobe of the labellum rounded, and 2(b) the mid-lobe of the labellum not rounded. 2(a) has two forms, *D. carinata*, or Bee Orchid, with yellow flowers very prominently marked red-brown, and *D. longifolia*, the Common Double tail, or Wallflowers *Diuris* (Plate 15), which occurs in all States except Queensland and the Northern Territory. It is a common southern species. The large wallflower-coloured flowers are hard to mistake. A variety with smaller flowers from Western Australia is var. *parviflora*.

2(b) has two species also, *D. emarginata*, or Tall Donkey Orchid, with a flower mainly yellow, and two basal ridges on the labellum lamina, and *D. pauciflora*, with yellow flowers spotted with brown and only one labellum lamina-ridge.

New South Wales has a large number of taxa at present under review. The main group of species of yellowish colours derive in the author's opinion from four sources: 1 *D. sulphurea*, 2 *D. maculata*, 3 *D. aurea*, and 4 *D. pedunculata*. These four seem to be reliably stable species widely distributed and with numerous forms. Many of these forms have been named as distinct species, but in reality most can be associated with one or another of the above four, or *D. platichila*. In the western plains area of New South Wales a large number of 'species' have been described, most of which are different shades of yellow with brown or red-brown markings. In the author's opinion many of these have originated from cross-fertilisation. If the flowers are dissected, flattened, and compared side by side the association to the parent species quickly becomes apparent. *D. sulphurea*, Sulphur Double Tail, which occurs in all States except the Northern Territory and Western Australia, has sulphur-yellow flowers with brown markings. The dorsal sepal is very broad with two dark brown spots at the base. The three-lobed labellum has the mid-lobe broad ovate, with no basal callus plates, but with a brown-red band or blotch on the upper surface—a very variable, spring flowering plant.

The Golden Diuris *D. aurea* (Plate 16), found in Queensland and New South Wales, is similar to *D. sulphurea*, differing in the colour of the flower shape of the labellum and by having two raised callus plates at the base of the labellum. It is very variable in colour and in the shape of the labellum mid-lobe, which varies from almost orbicular to ovate, the flower colour varies from golden-yellow to an orange-yellow with brownish marks. It flowers in the spring. *D. maculata*, or Leopard Diuris (Plate 17), occurs in all States except Western Australia and the Northern Territory. It is a distinct species, usually early-flowering, from July to September.

The Purple Diuris *D. punctata* (Plate 18) which occurs in Victoria, South Australia, New South Wales, and Queensland, is spring flowering. It is variable in colour and the length of the lateral sepals, which can be up to 9cm long. *D. palustris*, The Swamp Diuris, occurring in South Australia, Victoria, and Tasmania, flowers in the spring and is a small neat species, having many leaves, usually twisted. The flowers are small and fragrant, dingy yellow, boldly marked and spotted a dark red-brown. This taxon prefers wet swampy areas. *D. venosa* is a species confined in New South Wales to the Barrington Tops (back cover).

Diuris do cultivate fairly well in sandy soil.

Plate 17

Diuris maculata   LEOPARD DIURIS

All eastern and southern States. Spring flowering.

Plate 18

Diuris punctata   PURPLE DIURIS

The hover fly (*Eristalis tenax*) was not observed to act as a pollinating agent. Queensland, New South Wales. Victoria, and South Australia. Spring flowering.

Thelymitra SUN ORCHIDS

THE NAME THELYMITRA, from the Greek *Thelys*, feminine, and *mitra*, a turban or head-dress, refers to the hood of the column. It is a large genus containing about fifty species and many widely distributed forms, with approximately forty species, two varieties, and one form in Australia. Eleven species (six Australian types) and five varieties (one Australian) occur in New Zealand. Two (one Australian) occur in New Caledonia. Java and the Philippines have one species in common while one endemic species is found in Timor.

Sun orchids are terrestrial glabrous herbs, with two round or ovoid tubers and a solitary leaf, generally much elongated and fluted, linear to broad lance-shaped, oval-oblong, glabrous, or rarely hairy.

The flowers are at times singular, but more often in a terminal raceme. The colour is variable—

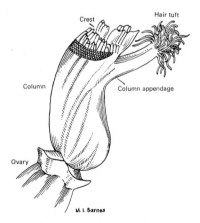

Thelymitra ixioides, side view of column and top of ovary. (After Scammel in Rupp's *Orchids of New South Wales*.)

white, pink, red, purple, blue, yellow, bicoloured, striped, and spotted flowers all occur. The flowers are almost regular, the labellum in some taxa only modified slightly, for instance *T. venosa*.

Thelymitra was described by J. R. and G. Forster[14] from plant material collected by them in the South Island of New Zealand during Captain Cook's second voyage. This species, *T. longifolia*, was also collected during Cook's first voyage by Sir Joseph Banks at Tolaga Bay in the North Island of New Zealand and was described by Dr Solander as *Serapias regularis*. It was illustrated in the Banksian plates, but was never fully described. Not a great deal has been written on the pollination of *Thelymitra*. R. FitzGerald included a few notes in *Australian Orchids*, but gives very little detail of pollinating agents.

Dr Rogers[15] appears to be the first adequately to describe the mechanics of the *Thelymitra* column. This genus relies on two methods of pollination— the use of an insect agent and self-pollination. Very little has been written on the insect agents. The cross-pollinating types usually have wide open flowers with united pollen masses attached to the viscid disc, whereas the self-pollinating types rarely open their flowers except in hot humid weather, and the pollinia are granular and not united to the viscid disc. These flowers pollinate themselves well before the bud is due to open. The one exception in this section is the variety *magnifica* of *T. venosa*, which opens at night as well as during the day.

Dr Rogers, using *T. luteocilium* as an example, described the self-pollination occurring in the following manner.

In early bud the column mid-lobe is cristate (crested or tufted), forming an incomplete hood. The anther occurs to the front and at the top of the column, while the blunt triangular point is between the lateral appendages. The pollen cases form two

Plate 19

Thelymitra media   TALL SUN ORCHID

Judging by their frequency in attendance, the hover flies *(Eristalis tenax)*, on this orchid could be possible pollinating agents. No confirmatory observation was made. Late spring.

Plate 20

Thelymitra ixioides   DOTTED SUN ORCHID

All States except Northern Territory. Usually blue with darker spots, but occurring in many colour shades. Spring.

long triangular bodies grooved between them. Their lower margins rest on the upper edge of the stigma while the rostellum is in the centre of the upper edge of the stigma plates. The column wings partly enclose the anther and pollen sacks to ensure that the pollinia fall directly on to the stigmatic surface.

At a more advanced stage, but still long before the flower opens, the pollen cases open by splitting vertically, and horizontally, at the base to expose the pollen grains. There is no connection between the pollinia and the rostellum.

The column meanwhile has increased in height, raising the anther higher, but the pollen cases still guard the pollen and its direction of fall on to the stigma. The fall of the pollen grains is brought about by the pull of gravity or the normal shaking of the flower in the wind.

In *T. fusco-lutea* (Plate 22) the process is similar except that the rostellum undergoes a degenerative process causing contractions, which assists the disposal of the pollen grains on to the stigma.

Dr Rogers states that *T. fuscolutea* is assisted in the separation of the pollen grains by a host of crawling insects that frequent this species. The insect was not identified.

In *T. antennifera* (Plate 24) Rogers describes the method used by a cross-pollinating species. Fairly early in the bud stage we find the column rather narrow; the anther point is broad, smooth and emarginate (indented). The column wings are not fully developed and the rostellum is very prominent. The pollen cases are vertical and the pollinia at their lower ends are in close contact with the rostellum.

As the column grows, taking the pollen cases with it, the wings widen and the stigma deepens. The rostellum conspicuously overhangs the stigma. By the time the flower is ready to open, the pollen cases have split; the column is now much higher and broader, having drawn up the pollen cases with it, causing an upward tug on the pollinia, which are well attached to the rostellum. The anther point has become roughened and curves forwards and downwards causing a greater pull on the pollinia, thus placing them in an almost horizontal position behind the stigma, where it would be impossible for self-pollination to occur. The pollinia in this position are easily removed along with the sticky rostellum by any intervening insect. The reddish-brown lateral appendages are used to attract the pollinating insects and to restrict the insect from going past the rostellum.

The plants also increase by tuber division; this occurs mainly in large matured plants; the tuber gradually wastes away at the centre of its mass until it divides into two separate tubers.

R. Brown divided the genus by the colour of the plants, and listed only ten species. G. Bentham, however, divides *Thelymitra* into three sections, using the column hood as the feature of distinction.

(1) *Cucullaria*. In this section the column wings are produced behind and beyond the anther into a broad hood over it, variously lobed or fringed at the top.

(2) *Macdonallia*. Here the column wings are broadly produced behind the anther, but much shorter and not hood-shaped.

(3) *Biaurella*. Here the column is not produced behind the anther. Instead it has two lateral erect lobes or appendages longer than the anther and often connected by a short crest behind it.

Group 1 contains most of our commoner species. The commonest would be *T. ixioides* (Plate 20) present in all States as well as New Zealand and New Caledonia, and flowering from August to December. Plate 20 shows this species in its pink form. Its colour is most variable, ranging from bright blue, pink, and (rarely) white, usually with petals and

30

Plate 21

Thelymitra venosa   VEINED SUN ORCHID

Generally of alpine occurrence in New South Wales, Victoria, and Tasmania, but also in marshy areas of South Australia. Summer.

Plate 22

Thelymitra fusco-lutea   BLOTCHED SUN ORCHID

Western Australia, South Australia, and western Victoria where they are often coastal. Late spring flowering.

dorsal sepal variously spotted black or a very deep blue. At times the spots can be absent. A number of varieties and forms are found in this species. Closely related to *T. ixioides* is *T. media* (Plate 19) of Victoria and New South Wales, which flowers from August to October, and differs in the column hood, the central tuft of which is missing, and in the fact that there are no spots on the segments. A pink variety *careno-lutea* (Victoria, and New South Wales) also occurs.

A beautiful rose-pink taxon is the rare *T. chasmogama* (Plate 26) of Tasmania, Victoria, New South Wales, and South Australia (September to October) its column wings fully cover the anther in a hood. Two spectacular plants are *T. fusco-lutea* (Plate 22) of Western Australia and Southern Australia, and *T. villosa* (Western Australia) which flowers from September to October. *T. fusco-lutea* is a fairly common plant in Western Australia and is found at its robust best in that State, its eastern point of distribution being at Wilson's Promontory, Victoria. This species appears to have originated in Western Australia where it is found in various types of habitat.

The column, with its wide, flared wings and its elaborate glandular process, is most unusual in this species. *T. villosa* resembles *T. fusco-lutea* closely, but is more robust in some forms. It has a very hairy ovate-oblong leaf, quite unique in the genus. In *T. villosa* we find that the lateral lobes— those that usually carry the cilia in other species— are modified radically into two conspicuous, glandular appendages decurved in front of the column. Another Western Australian species that has a similar colour scheme is *T. sargentii*. This differs from *T. villosa* in having a longer glabrous leaf and much shorter glandular appendages to the column; the rear of the column wings are more hooded.

T. aristata (Plate 27), is another cosmopoliton species occurring in all States except the Northern Territory, also in New Zealand (September to November). Commonly called the Scented Sun Orchid, this species is found growing on the north coast of New South Wales intermingled with *Dendrobium kingianum*, and has taken on a drooping habit much resembling the *Dendrobium*. The column hood is closed completely over the column and extends in front of it. The yellowish apex is bifid or notched in front. The hair tufts are white or pale pink resembling small toothbrushes. A pink variety *megcalyptra* with hood apex much inflated, is also known. *T. pauciflora* occurs in all States except the Northern Territory and also in New Zealand (August to December) and is a self-pollinating taxon closely resembling *T. aristata*, but usually smaller-flowered. The hood of the column is not as extended over the anther and the apex of the hood is variously deeply bifid. *T. pauciflora* has a wide range of colours, from white to deepest purple, mauve, pale blue, pink, and maroon in New Zealand. There is a deep violet-coloured variety *holmesii*, the segments often veined and the column hood longer than the type (South Western Victoria only).

The variety *pallida* with yellowish white flowers, white column hood and a yellow apex, is recorded from Bell, New South Wales. *T. nuda*, which occurs in all States except South Australia and the Northern Territory (September to November) is an ill-defined species intermediate between *T. aristata* and *T. pauciflora*. *T. longifolia* occurs in New South Wales, South Australia, Tasmania, New Zealand, and New Caledonia (October to December) and is a variable species in New Zealand. W. H. Nicholls does not include this species in his *Orchids of Australia* but Rupp[16] records a specimen from Sydney. Hatch[17] gives this taxon as divided into

Plate 23

Thelymitra rubra   SALMON SUN ORCHID

Widespread in Victoria and also present in Tasmania,
South Australia, and central coastal New South Wales.
Spring to early summer.

Plate 24

Thelymitra antennifera   RABBITS EARS

Confined to southern Australia—Western Australia,
South Australia, Victoria, and Tasmania. Late spring.

Photo: F. M. Coate

six varieties, one in South-east Australia and Tasmania; one in New Caledonia; one in the Auckland Islands, and three in New Zealand proper. He suggests the variety name of *australis* for the Australian variety. It is a slender plant up to 50cm high. The leaf is often but not invariably very long, variable in shape. Flowers are few, not large, generally blue or purplish with acute segments up to 13mm long. The column is hooded; the lateral lobes, small but with sparce hair, upturned (as in *T. pauciflora*). The column hood, usually very dark with yellow irregular edges, is tubular, extended well over the anthers.

Three species represent Group B. *T. rubra* (Plate 23) found in Tasmania, Victoria, South Australia, and New South Wales, is a slender species up to 50cm high. The flowers may be ruby-red, salmon, or pink, sometimes pale yellow with red marks. The column hood is not as high as the anther but shortly truncate, its margins denticulate (small-notched). The column appendages are about as high as the hood and covered with warts and glands.

R. FitzGerald first recognised this species from South Australia where it is often abundant, growing on heathlands and low hills. It is usually found in small tufts in groups of several plants suggesting a once common root stock. The variety *magnanthera*, known only from Jannali, New South Wales, has a very prominent anther.

A plant similar to *T. rubra* but smaller in height is *T. carnea*, which occurs in all States except the Northern Territory, also in New Zealand. The flowers are bright pink, salmon or creamy-white, about 1.5cm in diameter, rarely expanding fully.

The column appendages are smooth and not warted, and are about as high as the anthers. New Zealand has the variety *imberbis*, which differs only slightly from the type, being generally more robust.

The other species in this section is *T. flexuosa*, a yellow-flowered species which occurs in all States except Queensland and possibly New South Wales. A slender plant with wiry twisted stems and leaf, its flowers are small, up to four in number, lemon-yellow with red suffusions. The column wings are not hooded but slightly notched, much shorter than the anthers. The anther is produced into a large downy process. This species favours a damp habitat.

Group 3 has approximately seven representatives. The commonest taxon is *T. venosa* (Plate 21) found in Tasmania, Victoria, South Australia, New South Wales, New Zealand, and New Caledonia (November to February). *T. venosa* is divided into four named varieties, two, *venosa-venosa* and variety *cyanea*, are found in both Australia and New Zealand, the type extending into New Caledonia; variety *cedricsmithii* is found only in New Zealand and variety *magnifica*, a very large flowered form, is found only in New South Wales. The typical flower-type is a slender plant up to 75cm high. The flowers are blue, pink, or white, distinctively veined, with a smallish segment up to 18mm long, the labellum broader and with a wavy edge. The column wings are bright yellow with long twisted column appendages well above the anther. Variety *cyanea* is similar to the type but the column appendages are not higher than the hood and are curled in front of it.

The favoured habitat for this species is wet sphagnum moss bogs or on the margins of bogs in *Epacris* belts of usually wet, heathland.

One of the most attractive species in the group, *T. variegata* (Plate 25), occurs in Western Australia (September to October), a beautiful plant found on the coastal plains from Perth to Albany. Closely allied to this species is *T. spiralis*, found in Western Australia only (August to December) particularly the variety *punctata*. W. H. Nicholls[18,19] divides this

Plate 25

Thelymitra variegata   QUEEN OF SHEBA

Restricted to the coastal plains from Perth to Albany in
Western Australia. Spring flowering. *Photo: F. M. Coate*

Plate 26

Thelymitra chasmogama

Of rare occurrence in South Australia, western Victoria,
and Tasmania. Spring.   *Photo: P. A. Palmer*

species into four varieties—var. *scoulerae* with large, purple, flowers; var. *pallida*, with brownish green sepals, blotched, and petals pale pink; var. *punctata*, purplish, marked on the nerves with deeper spots; var. *pulchella*, pink, spotted chiefly on the sepals, deeper pink; var. *spiralis*, smaller than most of its varieties, purple or purplish-blue, lighter on the back of the segments with darker spots. *T. variegata* differs from the *T. spiralis* complex in its iridescent sheen and the much longer column appendages.

Another species closely allied to *T. variegata* is *T. mathewsii* of Victoria and New Zealand (September to October). A rare to very rare species in Australia, it has been collected in the Grampians area of Victoria in heavy scrub. A slender plant up to 22cm high, its leaf is broadly flared at its base then narrowing, spirally twisted. It has one or two dark purple flowers with darker veining; its column is erect, purplish at its base, changing to yellow at the apex; the midlobe is absent and the lateral lobes are dilated at their apex but connected at the base by a ridge of two calli.

A conspicuous taxon is *T. antennifera* (Plate 24) occurring in Tasmania, South Australia, Victoria, New South Wales and Western Australia (August to November). The common name 'Rabbits Ears' alludes to the conspicuous brown column appendages. It is a widespread species, having a rose-like perfume.

Epiblema BABE IN A CRADLE

THE GENERIC NAME *Epiblema* comes from *epi*, upon, and *bleme*, deformed, and refers to the column appendages of the orchid.

This is a monotypic genus, growing in very wet swampy conditions in Western Australia. At times it is even found in water, often in reeds or damp flats in conjunction with *Banksia* and *Callistemon.*

This genus was described by R. Brown in 1810. It resembles in habit the closely allied *Thelymitra* genus, but differs in the addition of a group of unusual appendages at the base of the column. The method of pollination is as yet unknown.

E. grandiflorum (Plate 28) is a slender plant up to 70cm high with a slender 30cm-long leaf. The flowers, numbering up to ten, are mauvish-spotted and veined with darker mauve. The labellum is broad ovate and in some cases the lamina is almost circular; the apex is rounded and prominently veined. At its base is a glandular process from which protrudes a number, from five to ten approximately, of stamen-like appendages. The column is short and squat and the stigma kidney-shaped. The column is framed by a conspicuous bifid column wing, which is higher than the anthers. It is found in Western Australia at Albany, Young's Siding, Spearwood, and Nomalup.

Plate 27

Thelymitra aristata   SCENTED SUN ORCHID

Widespread in all States except Northern Territory. Late
spring.

Plate 28

Epiblema grandiflorum   BABE-IN-A-CRADLE

Confined to swampy areas of coastal Western Australia.
Summer.   *Photo: F. M. Coate*

Prasophyllum LEEK ORCHIDS

THE NAME PRASOPHYLLUM, from the Greek *prason*, a leek, and *phyllon*, a leaf, refers to the plant's one stem-clasping leek-like leaf.

The first species of the genus were collected by Banks and Solander during Captain Cook's voyage along the eastern coast of Australia in 1770. The plant, *P. striatum*, was drawn by their artist Parkinson. The plate of this drawing was not made until 1780. This plant, along with twelve others, was described by R. Brown in 1810.

About eighty species occur within this mainly Australian genus. Five species occur in New Zealand, but two of these are of Australian origin. The plants vary in habitat from alpine meadows, which may be snow-covered for months, to semi-desert areas. Many of the species are difficult to identify as they closely resemble one another. Botanically they are divided into two sections.

(1) Euprasophyllum, in which the labellum is sessile, or on a rigid claw, and immobile in its connection to the column. Plants of this group favour damp and swampy areas and are often robust plants from 30cm to 1m in height.

(2) Micranthum, which is generally smaller, and often referred to as the Midge Orchids. The labellum is extremely mobile in all except one rare species, *P. transversum*. These plants usually grow in drier or sandstone areas and rarely exceed 30cm in height.

The pollination of this genus is usually effected by small insects—*P. australe* by Rove beetles, and *P. muelleri* by Chrysomelid beetles *(Ametalla spinolae)*. Ferment flies *Drosophila* (Plate 34), *Caviceps flavipes*, *Osinosoma subpilosa* and *Oscinosoma* Sp. have been observed to visit *P. morrisii*, *P. despectans*, and *P. archerii*.[20,21]

The plant prepares itself for the pollinating insect in the following manner. As a bud and freshly opened flower, the pollinia, with its attaching caudicle and viscid disc, lies against the rostellum in a special groove, the disc fastened to the top of the rostellum. As the flower ages the rostellum curves forward, tugging the caudicle and releasing the pollinia, thus allowing the drying air on to the strap-like caudicle, causing this member to curl and bring the pollen masses into a prominent position to be attached to any interfering insect.

The anther is often topped by what is called the anther point, an appendage-like structure which, in the case of the Midge Orchid, can also end with a gland. The apparent use of this structure is to restrict the flow of air on to the viscid disc and thus keep it in a viscid condition so it will readily adhere to the pollinating insect.

In a number of species in the section Micranthum, which rarely open their flowers, I have found all flowers produce fertile ovaries, suggesting self-pollination. The only plant in Euprasophyllum to exhibit any tendency toward self-pollination is *P. gracile*.

All parts of the flower appear to have their special uses. The labellum in the larger types appears to be used as a landing stage and the callus plate on the lamina acts as a guide to keep the insect on the right path to the stigma, whereas in the Midge Orchid the labellum, being hinged and very mobile, is used as a landing place and guide and also as an attraction to the insect. The column appendages appear to act as guides also, but in one case *(P. ruppii* var. *menaiense)* the column wings cross in front of the pollinated stigma, blocking further access to it. This only occurs after pollination.

Because the flower in *Prasophyllum* has the labellum uppermost it is said to be reversed, and according to some authorities goes through a 360-degree twist. But my observations have shown

Plate 29

Prasophyllum elatum   TALL LEEK ORCHID

Often exceeding 1m in height, this orchid occurs in all States except Northern Territory. Spring.

Plate 30

Prasophyllum suttonii   MAUVE LEEK ORCHID

An alpine, summer flowering plant, occurring in New South Wales, Victoria, and Tasmania, also New Zealand.

that in Prasophylla anyway this does not occur as the bud in its juvenile state is tightly pressed to the main stem and as it matures it rises and opens without any twist at all.

Prasophylla use three main attractions to induce the insect to visit them. The provision of easy access to the flower has been mentioned above; secondly they use a nectar secretion, and finally a scent. The presence of nectar can be clearly seen as small beads or droplets at the base of the labellum and column. The type of lure scent used depends on the insect which the flower needs to attract. In species like *P. odoratum* and *P. fuscum* it is a sweet pleasant perfume, but in some cases, it is a musky odour and in *P. striatum* it is a foul smell, possibly used to attract flies or blowflies.

In this genus we find some very unusual types of orchids. Two species deserving special mention are *P. anomalum* and *P. bowdenae*. In the former the column and in fact the whole flower is, truly anomalous, for the petals are not only fringed but also hinged, while in the column, the staminoides, which are usually fused together, are often disunited, giving the plant an appearance unlike an orchid. It can be found with the structure completely fused into a typical column. Two leaf bracts may be present with this species. The second plant, *P. bowdenae,* possesses much enlarged lateral sepals and labellum, resulting in the flower having an unbalanced appearance.

We find only one species that has developed a saprophytic trend and this is *P. flavum*, which has a much larger developed root system. Instead of the usual ovoid tuber it has a rhizome with very thick fleshy roots. Because the leaf has been reduced to a scale-like stem bract with the stem generally brown in colour, it suggests a reduction in chlorophyll.

R. Fitzgerald illustrated in *Australian Orchids* (1884) a plant under the name *P. australe*. His illustration does not represent this species, but an abnormal form of *P. flavum*. (Fitzgerald did not keep specimens of the plants he drew, so positive identification of some of his illustrations is difficult to arrive at with certainty).

Of the two sections Micranthum has the most concentrated distribution of the two as these orchids are found mainly on the central coast and adjacent slopes and tablelands of New South Wales. Thirty-five of the forty known species occur in this area, and only fifteen of these are found outside New South Wales. Of the other five, two occur in New Zealand *(P. nudum* and *P. pumilum)*, two, *P. brachystachyum* and *P. buftonianum*, in Tasmania, and the fifth *P. parvicallum* in southern Queensland.

Most of this section have minute flowers and one must use a hand lens to view the floral segments clearly. To the naked eye many of the species are very similar, but under the lens the various differences can be detected.

Three species that have been much confused in recent years are *P. brachystachyum*, *P. nigricans*, and *P. rufum*. These three are very closely related and may prove to be only one complex species. In Tasmania hybrids of the first two species do occur. *P. nigricans* occurs in Tasmania, Victoria, New South Wales, and South Australia, and has dark purple flowers with green markings, acute sepals and petals. Glands rarely present. The labellum has parallel sides for two-thirds its length, before acutely narrowing to a bristle-like tip, shorter than the lateral sepals, the column wings both acute and almost equal in length. The labellum margins are often minutely toothed.

P. rufum's labellum is not parallel-sided and has no bristle-like tip. The commonest plant in this section would be *P. archeri* (Plate 34), found in all States except Western Australia and the Northern Territory. It is a slender plant up to 35cm tall with

Plate 31

Prasophyllum gracile   GRACEFUL LEEK ORCHID

Recorded from South Australia, Victoria, New South Wales and Tasmania. Late spring flowering.

Plate 32

Prasophyllum brevilabre   SHORT-LIPPED LEEK ORCHID

Of scattered occurrence in all eastern States. Late spring and early summer.

the flowers (usually few in number), in a loose terminal spike moderately large for this section, greyish-green with purple markings. The dorsal sepal is broadly ovate, hooded, with lateral sepals narrower and a widely spread labellum, broadly ovate and often with an acute tip. The margins of the labellum and also the longer lobe of the column are densely fringed. Similar is *P. morrisii* (Plate 33), which occurs in Tasmania, Victoria, and New South Wales. It is a slender plant with a dense spike of hairy, purple flowers with hair on the dorsal sepal, petals, labellum, and column wings. A small appendage is present at the base of the stigma.

The Euprasophyllum section is not quite as puzzling to separate as the preceding group, but can still cause headaches. Dr Rogers[22] stated: 'The field observer who has paid special attention to this group rarely has difficulty in assigning any specimen to a particular species, but when it becomes necessary to record those salient points which definitely serve to distinguish one species from another his trouble begins; constant characters are not easy to find.' This statement still retains its truth after sixty years.

The commonest species is *P. elatum* (Plate 29), a plant of large dimensions up to 150cm, found in all States except the Northern Territory. Western Australia has a number of species with 'connate sepals' (the lateral sepals joined together with a thin membrane, which can at times be ruptured in hot dry weather) and most differ in labellum shape and reflexation of the lamina, as well as in the shape of the callus plate and its termination point. Perhaps

P. australe is the closest relation to *P. elatum*, differing only in the fact that the flower is more 'dumpy'. The labellum reflexes more, and has two knuckle-like swellings at its bend. *P. flavum*, occurring in Tasmania, Victoria, New South Wales, and Queensland, is similar to *P. elatum* but the labellum is more acute and the plant lacks a leaf. *P. brevilabre* (Plate 32) is found in Tasmania, Victoria, New South Wales, and Queensland and resembles *P. australe* in habit. But the flower spike is looser and the flowers are more slender. The labellum is narrower, the upper half reflexing right back on itself, giving the labellum a very short appearance, and the callus plate extends past the bend.

P. odoratum is the commonest of the 'sepals diverging' type found in Tasmania, Victoria, South Australia, and New South Wales. It has very fragrant flowers, the labellum resembling *P. suttonii*, but narrower. In *P. suttonii* (Plate 30) the flower's spike is usually short and rather dense; this plant is mainly alpine in habitat and in fresh flowers the lateral sepals can at times be connate. *P. patens* is an old species and at one time included *P. odoratum*. The labellum recurves back through the lateral sepals; the callus plate is inconspicuous, but it extends well past the bend. It is found in all States except Western Australia and the Northern Territory. *P. gracile* (Plate 31) occurs in Tasmania, South Australia, and New South Wales. A slender plant, the spike is many-flowered and the whole plant is yellowish-green. The labellum tip is constricted.

Plate 33

Prasophyllum morrisii   HAIRY LEEK ORCHID

Note the extremely hairy labellum. Occurs in Victoria,
New South Wales, and Tasmania. Autumn flowering.

Plate 34

Prasophyllum archerii   ARCHER'S LEEK ORCHID

The tiny fly *(Conioscinella becker)* having touched the
viscid disc with its back withdraws the pollinia on the
caudicle. The caudicle dries into such a position that when
the next flower is visited the pollinia will contact the
stigma, effecting pollination. Autumn flowering.

Chiloglottis BIRD ORCHIDS

CHILOGLOTTIS IS FROM *cheilos*, a lip, and *glossa*, a tongue, alluding to the shape of the labellum. It is a genus of about seven species, all Australian, with two species extending to New Zealand.

This genus divides easily into two groups, all species having only two leaves.

(1) Those with petals deflexed against the ovary. They are slender plants with the dorsal sepal much contracted in its lower half and the calli of the labellum in clusters.

(2) Those with spreading petals. The flowers are larger than those in (1) above, the dorsal sepal is not so contracted in its lower half, and the labellum calli are scattered.

Section 1 contains four species. The commonest one, the Autumn Bird Orchid, *C. reflexa* (Plate 35), occurs in Australia's eastern States. It can be identified by the large basal callus, which has a smooth stalk, and the head can be entire or bifid into two calli; smaller calli are most variable in arrangement. A very shy flowerer, *C. reflexa* grows in large colonies with few flowers. The flowering stem of this species and the whole genus elongates to disperse the ripe seed. The flowering period of *C. reflexa* can be almost any month of the year. The Broad-lipped Bird Orchid, *C. trapeziformis*, closely resembles *C. reflexa*, differing only in the trapeziform shape of the labellum and the single cluster of small glands in the labellum centre. Occurring in New South Wales and Victoria, it is common in some areas, and flowers from September to October.

Dockrill's Bird Orchid, *C. dockrillii*, occurs in New South Wales only and flowers in February.

It is a very rare plant from the Barrington Tops area of New South Wales, resembling *C. trapeziformis* generally but differing in the longer claw to the labellum and the dense labellum calli with two large jointed calli pointing towards the base of the column. The Ant Orchid, *C. formicifera*, closely resembles *C. reflexa*, differing mainly in the large labellum callus and in having its stem heavily warted; the top of this callus has two heads. The smaller calli are most variable in arrangement. The pollination of this species[23] and that of *C. trapeziformis* is effected by the ichneumon wasp *(Lissopimpla exselsa)* in a similar method as that described for the genus *Cryptostylis.*

Section 2 contains three species, the common Large Bird Orchid, *C. gunnii* (Plate 36), occurring in New South Wales, Victoria, and Tasmania. It flowers in September to February, rarely exceeds 6cm in flower, but elongates its fertilised ovary to 20cm for the dispersal of its seed. The labellum is broadly ovate with a short acute point, and the glands are very variable; it has a large stalked gland at the base and a shorter one in the centre of the labellum with two rows of very small glands either side of the main calli. Sometimes the glands are reduced to only three, giving the lamina a very bare look in comparison with the usual arrangement. The Green Bird orchid, *C. cornuta* (syn. *C. muelleri*), occurs in Tasmania, Victoria, New South Wales, and New Zealand, and differs from *C. gunnii* by having narrower leaves and a trowel-shaped labellum. Four calli form a central group with two more at the base, and this arrangement is variable. The Bronzy Bird Orchid, *C. pescottiana,* of Tasmania and Victoria, is very rare, and could even be extinct.

Plate 35

Chiloglottis reflexa   AUTUMN BIRD ORCHID

Chiefly autumn flowering, this orchid is present in all eastern States and Tasmania.

Plate 36

Chiloglottis gunnii   COMMON BIRD ORCHID

A plant of generally sheltered areas in New South Wales, Victoria, and Tasmania. Late spring to early summer flowering.

Caleana FLYING DUCK ORCHIDS

CALEANA HONOURS the name of an early botanical collector, G. Caley, and consists of five species.

Two, *C. major* and *C. minor*, are found in all States apart from Western Australia; *C. minor* is also found in New Zealand. *C. nigrita* occurs only in Western Australia, *C. nublingii* in New South Wales and *C. sullivanii* only in the Grampians area of Victoria.

The pollination[24] in this genus is by insect.

With the labellum in its normally upright position the insect—in the case of *C. major* a male saw fly, *Lophyrotoma leachii*—alights on the platform formed by the labellum lamina in a dorsal position. The labellum, being very irritable, springs itself back into the cavity formed by the column wings and the anther, taking the insect with it and either enclosing or forcing the insect between itself and the anther and viscid disc. The insect, dazed by the force of impact and momentarily stupified, is held still long enough for the viscid disc to dry on to the insect's thorax. The struggles of the insect plus the re-opening of the labellum enables the insect, laden with the pollinia, to escape. On visiting another flower of this species or genus the process is repeated. The attraction to the insect is not known, as the plant shows no sign of any secretion. The labellum, however, does resemble a small insect, particularly in *C. minor* and *C. nigrita*.

The labellum is very irritable in all species except *C. sullivanii*, where it is only slightly mobile. In this species the anther is abortive and it is doubtful if the species can reproduce normally from seed.

Caleana are found mainly in open forest country or open heathlands, often at the very base of large trees. The author has often wondered if some tree-orchid association occurs. Could the fungus assisting the orchid be common to both? The Large Duck orchid, *C. major* (Plate 37), flowers in spring and summer. It is a very slender plant up to 30cm high; the leaf is narrow and up to 12cm long, and red on the underside. Up to six green or dull reddish-brown flowers may be on the one stem, but there are usually less. The sepals and petals are inconspicuous, recurved in all species. The labellum is on a long irritable claw-like strap. The lamina is ovate with a beak-like projection resembling a duck's head. The column below the labellum is widely winged. The Small Duck orchid, *C. minor* (Plate 38) (November to January), is a much smaller plant, with flowers similar to *C. major* but smaller. The main differences are in the labellum, which is a deep red, sometimes green, and covered with dark glands. It has a bifid tip. The Western Australian Duck orchid, *C. nigrita*, flowers in October and is found in jarrah and red gum country. It resembles *C. minor* in habit but has a small ovate leaf at its base. The labellum is cigar-shaped and the glandular apex has a small appendage at the basal end. Nubling's Duck orchid, *C. nublingii*, is similar to *C. minor*, differing mainly in the almost pear-shaped, densely glandular covered labellum. Sullivan's Duck orchid, *C. sullivanii*, is similar to *C. minor* but the labellum shape is like a flat ice cream spoon. Its apex is raised and glandular.

C. minor is one member of this genus that responds well to cultivation and grows well in a compost of 60 per cent sand and 40 per cent leaf-mould. Moderate shade is required and it should be watered only when dry.

See article in *Contributions from NSW National Herbarium*, vol 4, no 5, p. 279, for reclassification of Genus Caleana and Spiculaea.

Plate 37

Caleana major FLYING DUCK ORCHID

All eastern States including Tasmania and South Australia. Late spring to early summer flowering.

Plate 38

Caleana minor SMALL DUCK ORCHID

Occurring in Queensland, New South Wales, Victoria, Tasmania, and South Australia, also New Zealand. Late spring to summer.

Drakaea HAMMER ORCHIDS

DRAKAEA IS NAMED in honour of the botanical artist Miss Drake. A genus of four Western Australian species, all the plants are very similar, differing mainly in the shape and ornamentation of the labellum.

The problems of pollination in this genus were first described by R. FitzGerald. He stated that he had not noticed any insects upon the plants he examined; in fact he had rarely noted a fertilised plant. He felt that insects were used when pollination occurred and believed that the genus could resort to self-pollination as a last resort. When it returned to the column before the flower faded, the labellum adhered to the rostellum and the action of the wind could remove the pollinia, brushing it on to the stigma. Mrs Ericson advances another solution, stating that as the labellum is so insect-like, a swooping insect hitting the labellum would be thrown back to the column where the pollinia would adhere to it.

All are slender plants having a heart-shaped leaf.

The Praying Virgin, *D. elastica*, which flowers from September to October, is the generic type plant. The labellum is short and upturned at its tip, the basal end thickly covered with hair and one tuft of glands. The Warty Hammer Orchid, *D. fitzgeraldii*, September-flowering, has very little hair on the labellum, the basal end of which is covered with glands. The King in his Carriage, *D. glyptodon* (Plate 39) flowers from September to October, and has the basal end covered with hair and a tuft of glands; the front tip is almost straight. The Hammer Orchid, *D. jeanensis*, is October-flowering, and has a labellum that resembles that of *D. glyptodon* but has no hair and a smaller basal tuft of glands.

Spiculaea ELBOW ORCHIDS

SPICULAEA IS DERIVED from the latin *spiculum*, meaning a sharp point or sting, and alludes to the very sharp-pointed flower or sting-like column. It is a small genus of three known Australian species— *S. ciliata*, endemic to Western Australia; *S. huntiana*, from the Victorian highlands and forest country and the New South Wales central tablelands and western slopes; and *S. irritabilis*, which can be found in areas of New South Wales and Queensland, extending to the Fly River area of Papua-New Guinea.

The genus was first described by John Lindley. Reichenbach united *Spiculaea* with *Drakaea* in 1858, but R. Schlechter in 1920 rightly separated the two genera. This peculiar genus and its taxonomic relations were discussed by Rogers[25] in 1926.

Rotherham was the first person to witness and photograph the pollination of *S. huntiana* in 1967. Pollination is effected by the pseudo-copulation of the male flower wasp of the Thynnidae family with the insectiform labellum[26] (Plate 40). *S. ciliata* has exhibited its resilience by opening flowers after two weeks under a compression of 80 pounds in a plant-drying press. This species has its flowers reversed to the other two species.

S. irritabilis, The Large Elbow Orchid, has a claw-hammer-shaped labellum with a hairlike tuft at the head. The apex is slightly glandular. The species produces two small elongate leaves connected underground to the leafless flowering stem. *S. huntiana*, Hunt's, or The Small Elbow Orchid (Plate 40) is a slender species bearing up to eight or more flowers. The labellum is insect-like in form, having a glandular head and body, and two long tail-like appendages. No leaf is known and the plant is strongly suspected of being a saprophyte.

Plate 39

Drakaea glyptodon

KING-IN-HIS-CARRIAGE
(HAMMER ORCHID)

Only of Western Australia occurrence.
Flowers in late spring and early summer.

Photo: H. A. Morrison

Plate 40

Spiculaea huntiana   ELBOW ORCHID

It was while photographing this orchid
that the pollinating agent and method of
pollination was observed. Pollination is
effected by the pseudo-copulation of the
male Thynad wasp (genus *Rhagigaster*)
with the orchid labellum.

Elythranthera ENAMEL ORCHIDS

ELYTHRANTHERA DERIVES FROM the Greek *elutron*, referring to the protective wing over the anthers, and the Latin *anthera*, meaning placed together.

The genus[27] was originally included in *Glossodia*, where it was erected as a new section by Endlicher in 1839 to cover the Western Australian species of *Glossodia*. G. Bentham retained Endlicher's section, but corrupted the name to *Eleutheranthera* and gave the eastern Australian species the section name *Euglossodia*. Bentham expressed the view that the two sections might almost be considered distinct genera.

While classified as *Glossodia* three species were recognised, but A. S. George rightly reduced *Glossodia intermedia* to a synonym of *E. emarginata*. I suggest that the pollination would be by insect agents, similar to some species of *Caladenia*.

It is a small genus of two species, both restricted to Western Australia. The Purple Enamel Orchid, *E. brunonis* (Plate 41), is widespread in the south-west of the State. It is most common in sandy soil. The glossy purple flowers vary in size and the tip of the strap-like labellum has one upward fold. The Pink Enamel Orchid, *E. emarginata* (syn. *G. intermedia*) is as widespread as the Purple species, preferring swampy soil of the coastal plains or the clay and granite soils of the forest areas. The large glossy pink flowers have a labellum similar to that of *E. brunonis* but it folds back, then forwards again in an S-shape. The tip of the labellum is notched, hence the specific name. White flowers occur in both species.

This genus does not appear to lend itself to cultivation, subsisting for a season or two before rotting. Much has to be learnt about how to grow many of the terrestrial orchid genera.

Adenochilus

ADENOCHILUS COMES FROM the Greek *aden*, a gland, and *kheilos*, a lip, alluding to the dense mat of glands on the labellum.

The genus consists of only two little-known species, one endemic to New South Wales and one to New Zealand. The Australian species is normally found growing in wet rock crevices or in sphagnum moss. It is a high-altitude plant rarely found below 3,000ft.

The *Adenochilus* genus is closely allied to both *Caladenia* and *Chiloglottis*.

The genus was described in 1855 by J. D. Hooker[28] using the New Zealand species *A. gracilis* as the generic type. *A. nortonii* was later described by R. FitzGerald in 1876, probably from Blue Mountain material. Little if anything is known regarding the pollination of this genus.

A. nortonii (Plate 42) flowers from November to December. It is a slender plant up to 20cm high. The flower stem which has a single leaf, rises separately from the stalked leaf, both coming from a creeping root stock. The leaf on the flowering stem is ovate to oblong-ovate, and the stem below the leaf is often spotted with red. Flowers are white, 1 to 2cm in diameter with a few short hairs on the edges of the floral segments. The labellum is shortly clawed with three lobes, the side lobes largish and erect while the middle lobe is smaller, reflexed, and naked. The centre of the labellum is densely covered with several rows of small stalked calli. The labellum and column are spotted with red. The labella calli are yellow. The column is slender and curved, with the column wings produced upwards, higher than the anther.

A. gracilis differs from the above in that the middle lobe of the labellum is covered with glands.

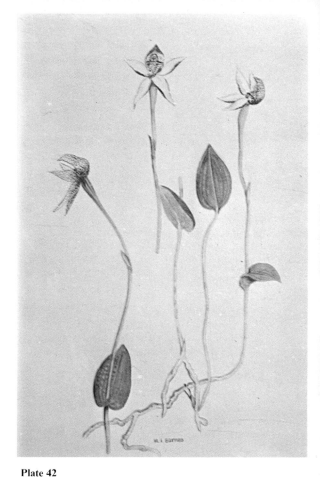

Plate 41

Elythranthera brunonis   PURPLE ENAMEL ORCHID

Both orchids of this genus are of Western Australia occurrence only. Spring.   *Photo: F. M. Coate*

Plate 42

Adenochilus nortonii

Confined to areas of the Blue Mountains and the Barrington Tops of New South Wales. Summer flowering. Drawing based on R. Fitzgeralds' illustration, Volume 1 *Orchids of Australia*, 1887.   *Drawing: M. I. Barnes*

Glossodia WAXLIP ORCHIDS

A SMALL eastern Australian genus of plants containing two species, *Glossodia* comes from the Greek *glossodes*, meaning tongue-shaped, and refers to the tongue-like appendage between the column and the labellum. This genus once included the Western Australian genus *Elythranthera*.

R. Brown described the two species of *Glossodia* in 1810 from material he collected 'in the vicinity of the colony of Port Jackson'. He distinguished *Glossodia* from its near relation *Caladenia* thus: 'An appendage between the labellum and the column. Another terminating the column which bears a membranous wing'. The appendage referred to is found at the very base of the column and is a large conspicuous callus approximately half the length of the labellum. It is solid with a double head in *G. major*, or divided into two in *G. minor*.

Pollination must be by insect agent, as the structure of the anther does not lend itself to self-pollination and known hybrids between *Glossodia* and *Caladenia* are found. This feature has been mentioned previously in literature, but to the author's knowledge the resultant plants (other than *Caladenia tutelata*) have not been described. At Nowra, New South Wales approximately ten plants of *Glossodia minor* and *Caladenia caerulea* have been located over several years. The plants generally resemble *G. minor*, although the colour often varies from blue-purple to magenta-purple. The plants grow up to 14cm tall and the flowers are 2 to 3cm broad. The leaf in some forms resembles *C. caerulea*, and is almost without hair; but in some specimens the leaf is typical of *Glossodia*. The sepals and petals are narrow as in *G. minor*, but the major points of

difference lie in the column and labellum. Some specimens have the labellum top-shaped, flat, with the side lobes not upcurved, or only slightly; no calli are present on the central disc of the labellum, but a few occur on the tip of the middle lobe. This lobe is white and conspicuous. In other specimens the labellum is typical of *C. caerulea*, complete with the two rows of golden calli. The column is closer to typical *Caladenia caerulea* but the wings are slightly longer. Both forms have the *Glossodia* appendage, and the red barred labellum and spotted column of the *Caladenia*.

Caladenia tutelata is considered to be a hybrid between *Glossodia major* and *Caladenia deformis*, or possibly *C. carnea*. This species is very rare and of infrequent occurrence.

Glossodia major, the Large Waxlip (Plate 43), is found in all States except Western Australia and the Northern Territory (August to October). Flowers number one to two, rarely three, up to 6cm in diameter, the rear of the segments lighter than the front. The Small Waxlip, *G. minor* (Plate 44), occurs in Victoria, New South Wales and Queensland (July to September). It is a slender plant similar to *G. major* in most respects, but smaller in all parts. The one to three flowers are deep violet-blue. The callus is divided into two sections. At times both species can be found with white flowers.

This genus does not lend itself to cultivation; the plants sometimes appear in the second year, but usually they rot during the dormant period. The habitat of *Glossodia* is varied; it often occurs in large number on open heathland in New South Wales, usually in a damp condition, in poor soil such as sandstone areas, or under the broken light of eucalypt forests.

Plate 43

Glossodia major   LARGE WAXLIP ORCHID

Flowers vary from deep magenta to white though blue-purple is the most frequent colour. All eastern States, Tasmania, and South Australia. Spring.

Plate 44

Glossodia minor   SMALL WAXLIP ORCHID

Ranging from far eastern Victoria, through New South Wales, to southern Queensland. Spring flowering.

Eriochilus BUNNY ORCHIDS

THE GENERIC NAME is from the Greek *erion*, meaning woolly, and *kheilos* a lip, alluding to the woolly labellum prominent in this genus. It is a small genus of approximately four species, one in the eastern States and South Australia, and three in Western Australia.

R. Brown noted the genus in 1810 when he described *E. autumnalis*, but this plant had already been described by the French botanist Labillardière as *Epipactis cucullata*. The three species of the genus from Western Australia were described by Lindley, who also named an *E. multiflorus*, which is now regarded as a robust form of *E. dilatatus*.

Mrs Ericson vividly describes the pollination of the genus by the bee *Hylacus dorothae*. The method described is as follows:

The stigma in *Eriochilus* is situated immediately below the anther and is in a deep cavity covered partly by two small flaps of the anther lid, and a fleshy flap which is the lower margin of the stigma. The pollinia, which are attached to the rostellum, occur under the two top flaps. This arrangement gives the plant no chance of self-pollination. The dense woolly labellum lamina provides an ideal foothold for the insect when delving for the sticky lower edge of the stigma. The bee that was observed pushed its head down, forcing it against the elastic-like lower flap, thus opening the stigma cavity as wide as possible. At this point, if the bee had previously visited a flower, the crest of pollinia on the bee's head is automatically inserted into the stigma. As it retreats the bee leaves the pollinia intact inside the cavity, where it is retained by the closing of the elastic-like lower edge of the stigma.

As the bee lifts its head it opens the two upper anther flaps, which shield the viscid disc from contacting the insect as it enters the flower, and allows the sticky disc to adhere to the still intact old viscid disc on the bee's head; thus by using this method the insect cannot pollinate the flower with its own pollinia.

Eriochilus flowers in the autumn, favouring damp places, but it can also be found on dry hillsides. The older the plant usually the larger the leaf and tuber. Normally the leaf is not fully developed when the flower opens in the autumn and continues to expand long after the flower fades.

Parson's Bands, *E. cucullatus* (Plate 46), occurs in all States apart from Western Australia and the Northern Territory (January to May).

Pink Bunny, *E. scabra*, occurs in Western Australia only, flowering from July to September. A fairly widespread plant up to 15cm tall, its flowers number one to three, and are pink with a velvet-red labellum. The dorsal sepal is broadly hooded at its apex, the lateral sepals broad, varying in colour from white to pink to rose-pink. The plants frequently grow close to creeks.

White Bunnies, *E. dilatatus* (Plate 45), occur in Western Australia only (March to June). The plant is usually slender, normally about 10 to 15cm tall, but can attain 30cm with up to fifteen flowers. (This is Lindley's *E. multiflorus*). The leaf is near the middle of the stem, just below the flowers. Flowers are white, the labellum marked red.

E. tenuis appears to be only a small-flowered form of the variable *E. dilatatus*, as the flowers are in all respects identical.

This genus appears to grow fairly well in cultivation, but does not multiply well.

Plate 45

Eriochilus dilatatus   WHITE BUNNIES

Confined to Western Australia, this plant, once known as
E. multiflorus, closely resembles the one species of the
eastern States. Autumn.   *Photo: F. M. Coate*

Plate 46

Eriochilus cucullatus   PARSON'S BANDS

The common name refers to oldfashioned clerical attire.
All eastern States from Queensland to Tasmania and
South Australia. Autumn flowering.

Caladenia SPIDER ORCHIDS

CALADENIA, FROM KALOS, meaning beautiful and *denia* gland, refers to the labellum glands.

The genus was described in 1810 by R. Brown with the description of fifteen species, but the genus now contains upwards of seventy species, the majority occurring in Australia. New Zealand has *C. lyallii* in common with Australia, plus three varieties and one form of the cosmopolition Australian *C. carnea*. In its variety *gigantea* this species extends to New Caledonia and the Malay Archipelago.

Bentham divides *Caladenia* into five sections, but with the addition of new taxa and new facts in relation to some species this is now reduced to four.

(1) Phlebochilus. Here the sepals are obscurely or distinctly acuminate, the dorsal sepal incurved and concave, erect behind the column or reflexed with it. The labellum is broad, with deeply coloured diverging veins either undivided or with a very small and obscure mid-lobe. Examples: *C. cairnsiana, C. multiclavia.*

(2) Calonema. Sepals are acuminate, with long or short points, with dorsal sepals erect and incurved. The labellum is inconspicuously veined, the lamina with two or more rows of calli. Examples: *C. filimentosa, C. patersonii.*

(3) Eucaladenia. Sepals are acute or obscurely acuminate, rarely obtuse, the dorsal sepal usually erect and concave. The labellum is inconspicuously veined, and the lamina has two or more rows of calli (sometimes arranged or united at the base, almost in a semicircle). Examples: *C. carnea, C. flava* and *C. testacea.*

(4) Dentisia. Sepals and petals are obtuse and nearly equal, all spreading. The labellum and column are very short, the calli small, and numerous in longitudinal rows. Example: *C. gemmata.*

Within Australia the taxa found in the eastern States are divided between sections 2 and 3, the former being referred to as Spider Orchids, alluding to the long filimented sepals giving these plants a 'spidery' appearance. The latter are often referred to as the 'carnea' group, as *C. carnea* is the commonest species.

Sections 1 and 4 are confined to Western Australia, as are the majority of section B, this State having the largest number of these beautiful plants, forty-four species.

Within the genus a great range of size occurs, from the diminutive *C. carnea* var. *pygmaea,* only 3 or 4cm high, to giant forms of *C. patersonii* var. *longicauda*, 80cm high and 30cm wide, with sepals out-stretched.

The cross-pollination of Caladenia has been discussed by Dr Rogers[29] in 1931 where he investigates *C. deformis*. An earlier Western Australian observer, O. Sargent[30] describes the pollination of *C. barbarossae*. R. FitzGerald[31] makes only three brief notes on the subject. Rica Ericson[13] describes admirably the various contrivances used by Western Australian *Caladenia* to secure pollination.

O. Sargent conducted his observations in 1905 and 1906 on plants growing in the bed of the Avon River, where they are covered for weeks by floodwaters. They bloom in late spring after the floods subside. Sargent noted the pollination of *C. barbarossae* by an unidentified wasp. In this species the labellum is mobile on an exceptionally long column foot; the lamina has a group of insectivore-shaped glands at its base and is fairly densely covered with cilia. The wasp alights on the lamina and proceeds towards the group of glands at the base, thus causing the lamina to topple over, taking the wasp with it and throwing it against the column. As the wasp backs out the pollinia are removed with it as the lamina tilts back to its original posture.

Plate 47

Caladenia patersonii
COMMON SPIDER ORCHID

The blowfly, apparently being eaten by the orchid, helped establish the common name for this orchid genus. The fly is 'ensnared' during pollination procedures (see text). All Australian States except Northern Territory. Spring flowering.

Photo: H. A. Morrison

Plate 48

Caladenia flava   COWSLIP ORCHID

Only of Western Australia occurrence but widespread in that State. Spring.

Photo: H. A. Morrison

This method is fairly typical of these plants which have a movable labellum (see Plates 47, 50).

The method used by plants of section 3 is described by Dr Rogers. This method was brought to the doctor's notice by Mr H. Goldsack, a prominent South Australian collector, who observing that a bee had several pollinia adhering to its back, collected it as well as the plant. On opening the box the following morning he found the insect wedged tightly in the tubular space formed between the column and labellum of the orchid. When disturbed it hurriedly backed out on to the free extremity of the lip. During this movement the pollinia were rubbed hard against the stigma. Later the bee returned to 'repast', and it was possible to follow its subsequent movements more carefully. It was able, without any exertion, to penetrate the space referred to until a point was reached when the mesothorax was on a level with the stigma. Then in its effort to reach the calli at the base of the labellum, it pushed the latter away from the column, at the same time exerting strong pressure with its back against this structure and incidentally pressing the pollinia on to the stigma.

The reference to the bee's back contacting the stigma is most important, for with *Caladenia* the pollen masses contain no sticky viscid disc or caudicles of any kind but being rather granular, the pollinia adhere to anything sticky, such as the bee's back. The bee's identity in Dr Roger's example was *Halictus subinclinans*.

In the case of *C. latifolia* pollination has been observed enacted by a pollen-eating beetle.

Often with orchids in section 2 a fly or small bee is found dead at the base of the column, giving rise in some places to conjecture that the so-called Spider Orchid eats the trapped insect. This is not the case. The reason for the death of the trapped insect is its inability to back out of the cage formed by the lamina and column. This is nature's way of making sure that only the right-sized insect can remove the pollinia. (Plate 47.)

Nobody as yet seems to have been able to understand the attraction *Caladenia* offers the pollinating insect. Mrs Coleman suspected it could have been a 'sex lure', as she observed the ichneumon fly pollinating *C. dilatata* var. *rhomboidiformis* in a similar manner to *Cryptostylis*. Dr Rogers stated that no nectar was present at the base of the labellum of *C. deformis*, but it is present in many species.

Habitat of the genus is wide and varied—in the sandy loam of the seashore, in clay soils of the open forest country, in dry ridge areas, open swampy heathlands, in damp areas, alpine meadows, and semi-desert areas. While *Caladenia* are mainly spring flowering, some species bloom in summer and autumn or late winter.

The description of some of the commoner species in each section follows.

Section 1 contains only Western Australian species and does run somewhat into section 2. Possibly the commonest would be *C. discoidea*, Dancing Orchid, a widespread species up to 30cm high. It often has three flowers of blunt appearance, yellowish-green with red markings and a broad, fringed, labellum with distinct dark red veining, the crowded calli irregular. The rarest of this section is *C. multiclava* the Lazy Spider Orchid (Plate 49), a very peculiar plant; the column is at 90 degrees to the ovary, thus placing the labellum immediately above the ovary. The labellum calli are joined together to form a raised matt.

Section 2 is by far the commonest and most conspicuous of the genus. Of the thirty-seven species and seven varieties in this section Western Australia has twenty-four species and four varieties. The White Spider, *C. patersonii* (Plate 47), is very widely distributed and in its type form is found in

Plate 49

Caladenia multiclava   LAZY SPIDER ORCHID

Confined to Western Australia, this uncommon orchid
has the labellum horizontally positioned at right angles
to the ovary. Spring.   *Photo: H. A. Morrison*

Plate 50

Caladenia dilatata   GREEN-COMB SPIDER ORCHID

The Thynid wasp *(Phymatothynnus monilicornis)* has
been tipped into the flower after attempting to mate with
the hinged labellum. The pollinia on the wasp's back can
be seen just prior to being deposited on the stigma—on
backing out of the flower another pollen mass will be
attached to the insect.

all States except the Northern Territory. Plants of the var. *longicauda* are some of the loveliest of our ground orchids. Two other varieties of this species are known, var. *arenaria* with light grey flowers and with the labellum marginal fringe crenulate only, and the var. *concolor* (Plate 2) which is prune-coloured. The labellum margins are less fringed than in the type, but not as short as var. *arenaria*. Both varieties occur in Victoria and New South Wales. Sepals in this species are non-clavate. The Butterfly Orchid, *C. lobata*, is one of the most beautiful of all the Spider Orchids, being pale greenish-yellow with crimson markings. The labellum which is rather large and delightfully coloured, poised delicately on the very broad claw and trembles at the slightest zephyr, resembling a butterfly. It occurs in Western Australia only, flowering in the spring.

The Green Spider, *C. dilatata* (Plate 50), is found in all States except Queensland and the Northern Territory. A widespread but not usually common plant, it is found alone or in small groups. The green flowers with the distinctive maroon tip to the labellum prevail through the three varieties. Var. *falcata* of Western Australia has the lateral sepals bent upwards abruptly, while the other variety *rhomboidiformis* found only in that State, has a rhomboidal-shaped labellum and very short marginal combs. The variety, *concinna*, is like a small form of the type, but the labellum fringe is reduced to mere teeth. The type form is usually clubbed on its sepals, but non-clubbed forms are known.

The Red Spider, *C. filimentosa*, is another widely distributed plant occurring in all States apart from Queensland and Northern Territory. It is easily distinguished by its extremely long filimented sepals and petals. Its general colour is maroon or red-brown with lighter markings; its var. *tentaculata*, Yellow Spider, is a cream or yellow colour form with reddish marks and occurs in Victoria, South Australia, and Western Australia.

Section 3 is widely represented in all States and contains twenty-five species, ten varieties, and one form. The commonest species is *C. carnea* (Plate 51) found in all States except Western Australia and the Northern Territory and extending to New Zealand and Java. In its type form it is a slender plant but var. *pygmaea* and var. *minor* are very diminutive, whereas var. *gigantea* can reach 50cm in height, with bright pink flowers 5 or 6cm across. Plants of this species can have a delightful honey scent, but others lack it. All forms placed into *C. carnea* must show the bright reddish bars on the labellum and column. *C. alba* is a plant very closely related to *C. carnea* and where these two mingle, hybrids occur. They are usually identified by the labellum and the labellum callus-shape, the column shape, and height of the wings. *C. alba* var. *picta* has a tri-coloured column, but recent observations have shown this variety to run into the type form and should possibly be included under the species. The Blue Fairies, *C. deformis* (Plate 52) are a common species of the southern States, which extends to all States except Queensland and the Northern Territory. It is found in rare cases in white and yellow. *C. flava* (Plate 48) is a prominent Western Australian species; its brilliant yellow colour with red marks needs no other guide to identify it. Pink Fairies, *C. latifolia* (Plate 53) are common in some areas, and occur in all States except the Northern Territory, with the Queensland record rather doubtful. They are rare in New South Wales. In Western Australia this species reaches its best with broad long leaves, normally 6cm x 1.5cm wide, but the author has examined plants with leaves 20cm long x 1.75cm wide, possibly a shade form or a plant that grew in long grass.

Section 3 can be divided into two groups. The

Plate 51

Caladenia carnea   PINK FINGERS

An abundant species with many forms. In all States except Western Australia, it extends to the Malay Archipelago and New Zealand.

Plate 52

Caladenia deformis   BLUE FAIRIES

Mostly of coastal occurrence in all States except Queensland and Northern Territory. Spring.

above represents the 'carnea' group; we shall now discuss what is commonly referred to as the 'angustata' complex. In 1931 W. H. Nicholls[32] figured and discussed section 3 in the light of the then known facts, but there is still much to be sorted out in this latter group. The Musky Caladenia, *C. angustata*, occurs in Tasmania, Victoria, Southern Australia and New South Wales and is a slender plant, the largest in the complex. The flowers are white inside, the outside being chocolate and very glandular; the labellum is three-lobed and streaked with purple. Smaller than this species is the Hooded Caladenia, *C. cucullata*, which occurs in Tasmania, Victoria, and New South Wales. It is a distinct plant; the upper sepals and column are decidedly bent forwards; the labellum, tipped with bright purple, has dense calli, and the marginal calli have their stalks glandularly hairy (seen under a x10 lens). *C. testacea* is a New South Wales plant that extends into Victoria, where it is rare. It differs from the former in its dorsal sepal and column, which are curved; the labellum calli are not as crowded and the marginal ones not as hairy. *C. gracilis* from Tasmania is similar but smaller than *C. angustata;* the outside of the segments are pink-brown. Also resembling *C. angustata* is the rare *C. iridescens*. This plant with its golden iridescent hue is hard to mistake. This section would

not be complete without a mention of *C. menziesii* (Plate 54) the Hare Orchid, which occurs in Tasmania, Victoria, Southern Australia and Western Australia. This plant differs materially from the rest of the genus and at one time this species and the closely related *Leptoceras fimbriatum* were the bases of G. Bentham's section *Leptoceras* of *Caladenia*. It is a slender species up to 24cm high, the leaf smooth and broad, with up to three white and pink or occasionally all-white flowers. The petals are conspicuous, like two long 'hare's ears'. This species is usually abundant after bushfires as are many of the Australian ground orchids.

Section 4 contains only one Western Australian species and its form, *C. gemmata*, China Blue Orchid, which flowers from August to October. It is a short plant with an oval leaf (unusual for a *Caladenia*) about 2.5cm long, dark brown below the upper surface, and covered with short bristle-like glands. The large dark blue flowers (rarely white) have broad obtuse segments. The unusual labellum is broadly oval, the margins are entire and the calli studded over the lamina are short and clubbed with two much larger calli at the base. The inland forms of this species are often darker in colour and the flowers larger than the coastal forms. The form *lutea* has a smaller all-yellow flower. Recently a white form of this species was found.

Plate 53

Caladenia latifolia   PINK FAIRIES

An abundant coastal orchid of all States except Queensland and Northern Territory, although rare in New South Wales. Spring.

Plate 54

Caladenia menziesii   HARE ORCHID

Flowers well after fires and occurs in Western Australia, South Australia, Victoria, and Tasmania. Late spring flowering.

Burnettia LIZARD ORCHID

THE EARLY botanical collector Burnett was honoured by John Lindley when he recorded this genus in 1840. It is an extremely elusive orchid in most of its distribution—though widespread it is very rarely seen in flower. Very few collections have been made in New South Wales, where it has been recorded in coastal swamp and in alpine and tableland areas. In Victoria and Tasmania the plant appears to favour boggy and marshy areas often associated with dense *melaleuca* growth. The species is most often seen in flower after bushfires. The removal by fire of the dense cover under which these plants usually grow makes their detection easier. The draining of swamps and marshes and the clearing of the plant's habitat for cultivation is reducing known habitat areas considerably.

While the popular name Lizard Orchid is generally accepted over the plant's Australian range, this name should not be confused with the Lizard Orchid of Europe and England, *Himantoglossum hircinum*, or the North American orchid that shares this common name.

Burnettia cuneata (Plate 55) of Tasmania, Victoria, and New South Wales is September–November flowering. A small plant up to 10cm with no leaf present at the time of flowering it usually has two to three flowers, white with reddish or purplish-brown markings on the back of the petals and sepals. The two side-lobes of the labellum roll inwards, tending to give the labellum a tubular appearance and it is often marked with reddish lines. A patch of calli occurs towards the apex of the labellum and these extend to the base as two callus plates.

Leptoceras
FRINGED HARE ORCHID

LEPTOCERAS IS DERIVED from the Greek *leptos*, meaning delicate or thin, and *keras* a horn, referring to the two delicate hornlike petals. This is a monotypic genus occurring only in Australia.

The plant flowers most profusely after fires have burnt over the areas where it occurs. Either potash or some ingredients of the ash appear to induce flowering.

Fringed Hare Orchids favour sandy areas, often establishing huge colonies, but unless fires have occurred few plants flower.

No record is known regarding pollination in *Leptoceras* but it would almost certainly be similar to that of the *Caladenia* genus, to which this plant is closely allied.

The Fringed Hare Orchid *L. fimbriatum* (Plate 56) is autumn flowering and occurs in Victoria, South Australia and Western Australia. It is a slender plant attaining 20cm in height, usually with one broad lance-shaped leaf. The leaf is absent or very small at flowering time, but as the plant matures the leaf increases in size and the parallel veining becomes distinctly marked. It has one to three flowers, generally reddish or yellowish-brown, with erect petals like the ears of a hare. The lateral sepals are narrow and deflexed and the dorsal sepal is concave over the column. The labellum is obscurely three-lobed with a distinct greenish-red fringe and a few red patches on the lamina. The flowers tend to persist on the plant long after the seed has ripened.

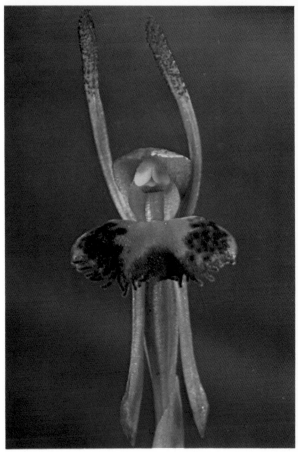

Plate 55

Burnettia cuneata   LIZARD ORCHID

Uncommon to rare in its Tasmania, Victoria, and New South Wales occurrences. Spring flowering.

Plate 56

Leptoceras fimbriatum   FRINGED HARE ORCHID

Ranging from Western Australia through South Australia into western Victoria. Autumn.   *Photo: J. B. Fanning*

Rimacola GREEN ROCK ORCHID

A MONOTYPIC GENUS restricted to the central and southern coastal areas and adjacent tablelands of New South Wales, *Rimacola* occurs in damp sandstone crevices. The name *Rimacola* means 'inhabiting crevices'.

The plant is generally found growing on sandstone cliffs either in the roots of *Todea barbara* (king fern) or *Gleichenia circinata* (coral fern), or other damp crevices, usually in a very dense mass with water dripping through the roots.

It was originally described by R. Brown in 1810 as *Lyperanthus ellipticus*, although it hardly resembles any of that genus. Reichenbach removed it to *Caladenia*, but this treatment was not accepted. In 1942 the Rev. Rupp[33] created a new genus *Rimacola* to accommodate the anomalous *Lyperanthus*.

The old flower-spikes in this genus are very persistent, often being found on the plant for a number of years after fading.

The pollination of this genus has not been studied, but it is certainly effected by insects, as the number of pressings in the author's herbarium only show three ripened capsules in thirty-eight flowers.

Green Rock Orchid, *R. elliptica* (Plate 57), occurs only in New South Wales and flowers in the late spring or summer. It is a slender plant of about 20cm with a succulent rhizome as the root stock. The elliptical leaves occur as a rosette or are stem-clasping and the flowers are in a rather crowded drooping raceme. The labellum is white with red streaks and splashes. All segments of the flower are slender and filiform. The labellum is set on a claw and is entire with a few or no calli. The slim column is narrowly winged above.

Right: Meadow Argus butterfly—possible pollinating agent of *S. sinensis*. Photo: R. Badgery.

Spiranthes LADIES TRESSES

SPIRANTHES IS a large cosmopoliton genus with upwards of 300 species widely dispersed throughout the temperate zones of both hemispheres, from Siberia through tropical Asia to Australia, Tasmania, and New Zealand. *Spiranthes*, from the Greek *speira*, meaning a coil and *anthos*, a flower, refers to the spiral arrangement of the flowers.

The genus in Australia is represented by only one species (Plate 58) which extends from many stations in Asia through China and Malaya to all Australian States apart from Western Australia, and also New Zealand.

In its northern stations (Asia etc.) the plant has flowers that open fairly widely and are insect-pollinated, but as the species progresses south the flowers tend to become closed and tubular and in New Zealand become self-pollinating.

One instance of possible pollination is shown below being carried out by the Meadow Argus butterfly *(Precis villida calybe)*. Mrs Coleman gives an excellent description of the method used by *S. sinensis* for pollination. The old flower is pollinated by freshly-opened flower pollinia, as the younger flower releases its pollinia before the column wings dilate in the older flowers to bring the stigma into a receptive state. The stigma in the young flower is not viscid.

Plate 57

Rimacola eliptica GREEN ROCK ORCHID

This rare orchid is confined to wet crevices in the sandstone of the Blue Mountains and adjacent areas of New South Wales. Late spring to summer. *Photo : E. Gordon*

Plate 58

Spiranthes sinensis AUSTRAL LADY'S TRESSES

Usually in marshy habitats of all States except Western Australia, also New Zealand and Asia. Summer.

Calochilus THE BEARDIE

A SMALL GENUS of approximately eleven species chiefly Australian but extending to New Zealand, New Caledonia, and Papua. *Calochilus* is derived from the Greek *kalos*, meaning beautiful and *kheilos*, a lip, alluding to the beautiful bearded labellum in most species.

Calochilus was described by R. Brown in 1810, the generic type being *C. campestris* and *C. paludosus*. Lindley in 1840 added *C. herbaceus* and in 1873 Bentham added *C. robertsonii*. *C. holtzei*, from the Northern Territory, was described by Baron von Mueller in 1892. The New Caledonian representative of the genus, *C. neocaladenicus*, was described in 1907 by R. Schlechter. Rogers added *C. imberbis* in 1927, with the very rare *C. richae* being described by Nicholls in 1928. Rupp described the beautiful *C. grandiflorus* in 1934 and again in 1943 he added *C. gracillimus*. The Papuan *C. caeruleus* was added by L. O. Williams in 1946.

Description of the pollination of the genus is restricted to an article by F. Fordham in 1946, where he gives data on the pollination of *C. campestris* by the Scollid wasp, *Campsomeris tasmaniensis*. This plant is also self-pollinating.

Copper Beards, *C. campestris* (Plate 60), occur in all States except Western Australia and the Northern Territory. They also extend to New Zealand (September to November). The plant illustrated as this species by R. Fitzgerald in *Australian Orchids* is in fact a pale-flowered form of *C. robertsonii*. A robust plant up to 70cm tall, its leaf is thick, up to 24cm high and deeply channelled. Flowers number up to fifteen and are comparatively small, yellowish-green with reddish brown or brownish markings. The labellum has a fleshy naked rectangular base, extending into a wide triangular densely hairy lamina. The extremities of the labellum extend to a ribbon-like appendage. The column has a black gland on each side at its base.

The rare Pale Beard Orchid *C. herbaceus*[44], occurring in Tasmania, Victoria, and New South Wales, resembles *C. campestris* but is paler in colour and lacks the robust leaf. It has a much larger root system of thick fleshy roots as well as odd-shaped tubers. More robust than *C. campestris* with up to twenty flowers, the petals are as long as the sepals; the apex of the labellum is naked and triangular below the hairy lamina. The column lacks the black glands as in *C. paludosus, C. richae* (frontispiece) is extremely rare, found only in Victoria; it differs from all other species in that the labellum lamina is almost circular. The labellum lacks hair, these being replaced by very short glands. First found in a nearby locality to *C. richae* is *C. imberbis* (back jacket), and the labellum of *C. imberbis* is ovate acute, the tip somewhat rolled inwards. The Great Beardie, *C. grandiflorus* which occurs in Queensland, New South Wales and north-eastern Victoria, is a slender plant up to 60cm tall. The flowers, numbering up to ten, are large for the genus and very similar to *C. robertsonii*. The flower colour is golden-yellow and the purple labellum is broad and glandular at the base, as with *C. robertsonii*. The column glands are not connected with a colour ridge. *C. gracillimus* of New South Wales and Southern Queensland (November to January) is closely allied to *C. robertsonii* (Plate 59) but has a long slender labellum. *C. robertsonii* occurs in all States and in New Zealand (September to January). The labellum hair is very fine. *C. paludosus* has no glands at the base of the column and occurs in all States except Western Australia, also New Zealand.

Plate 59

Calochilus robertsonii   BROWN BEARDS

All States except Northern Territory. Also found in New Zealand. Spring.

Plate 60

Calochilus campestris   COPPER BEARDS

All eastern and southern States except Western Australia, also New Zealand. Late spring to early summer. (See also Frontispiece and book jacket.)

Lyperanthus BEAK ORCHID

THE GENERIC NAME *Lyperanthus* derives from *Lyperos*, the Greek for mournful, and alludes to the type species, which turns black when pressed.

This is a small genus of about nine species, four of which are endemic to Australia, the other species extending to New Zealand and New Caledonia. R. Brown first described it in 1810, listing three species, *L. suaveolens*, *L. ellipticus*, and *L. nigricans*. Lindley later added the Western Australian *L. serratus* and finally Baron von Mueller described the rare Western Australian plant *L. forrestii*. Of the above *L. ellipticus* has been renamed *Rimacola ellipticus*.

Many populations of *Lyperanthus* only flower in profusion after a bushfire. It is difficult to know if it is the heat of the fire that causes this, or the added amount of potash that is left after the burning of the leaves and forest debris.

The normal habitat of the genus is variable, for it can be collected in dry heathland or swampy, sandy areas. The plants are often seen in extensive colonies in these habitats, but other than *L. suaveolens*, a plant that flowers regularly, few flowers are seen in a normal season.

Little is known about the pollination of the genus, although it is thought that insects act as the pollinating agent.

Brown Beaks, *L. suaveolens* (Plate 62) is fairly common in heathland country in the eastern States. Its sweet-scented flowers have a vanilla-like perfume and are freely visited by small native bees.

It is often difficult to locate, as its colour blends well with its natural habitat. When not in flower the leaves resemble some of the wide-bladed grasses. It can attain a height of 44cm but is usually 20–30cm high. As with all this genus it is spring flowering, and occurs in Tasmania, Victoria, New South Wales and Queensland.

Rattle Beaks, *L. serratus* is the Western Australian counterpart of *L. suaveolens*, which it resembles fairly closely. It has a long narrow-bladed leaf, but the flowers are more robust and have a larger sheathing floral bract. The flowers are yellow-green with dark crimson streaks and have a large group of yellowish glands forming the mid-lobe of the labellum.

Apparently Brown named the genus from Red Beaks, *L. nigricans* (Plate 61), as it turns jet-black when dried. This is caused by the indigo pigment present in the plant. It has a similar habitat to the preceding species. The plants, with their large heart-shaped, fleshy leaves, are found in colonies of from 3 to 100 or more plants, occurring in all States apart from Queensland. Forrest's Beak Orchid, *L. forrestii*, is by far the rarest of the genus in Australia, being found only in Western Australia, in the Sterling Ranges and south to the coast.

It resembles a small *L. nigricans* but with three smaller and longer basal leaves, usually with only two flowers which are more upright and open and a lighter colour. The flowers are heavily perfumed. The labellum is not as fringed at the mid-lobe.

Lyperanthus in general do not respond well to cultivation, *L. suaveolens* probably being the most amenable.

Plate 61

Lyperanthus nigricans   RED BEAKS

All States except Queensland and Northern Territory.
Spring.

Plate 62

Lyperanthus suaveolens   BROWN BEAKS

Occurring only in New South Wales, Victoria, and
Tasmania. Spring flowering.

Acianthus MAYFLIES,
PIXIE CAPS AND GNATS

THE GENERIC NAME *Acianthus* comes from the
Greek *akis* a point and *anthos* a flower, alluding to
the pointed appearance of the flower. It is a genus
of about twenty species. Seven species occur in
Australia, the varieties of two of them extending
into New Zealand. Twelve more taxa occur in
New Caledonia. It is a variable genus in Australia,
favouring sheltered eucalypt forest, under bracken
fern *(Pteridium esculentum)* or behind the sand-
dunes on the beach front under *Leptospermum* in
the grey sandy leaf-mould.

This genus was described in 1810 by R. Brown
as numbering three species, *A. caudatus, A. fornicatus* and *A. exsertus*. Brown also described
Cyrtostylis reniformis, but this genus was included
into *Acianthus* by Schlechter in 1906.

In 1891 Bailey described a plant from Queensland
under the name of *Microstylis amplexicaulis* in his
Queensland Flora of 1902. Bailey classified this
same species as a *Listera*. Subsequently in 1903 the
plant was included in *Acianthus* by Rolfe.

From Western Australia comes *A. huegelii*. In
1938 Rupp described the south Queensland species
A. ledwardii.

A. amplexicaulis of Queensland and northern
New South Wales flowers in April. It is a slender
plant up to 12cm tall with a sharp-pointed, prominently veined, heart-shaped leaf. The small flowers,
up to seven, are supported by a large floral bract.
The sepals and petals are narrow and the labellum
is trapeziform and decurved at almost 90 degrees
in the centre. The apex of the lamina is somewhat
toothed. The Mayfly Orchid, *A. caudatus* (Plate
63), occurs in Tasmania, Victoria, New South Wales
and South Australia. It flowers in September and
October, has no parallel in the Australian species

of the genus. This taxon produces an unpleasant
scent. From Tasmania and New South Wales comes
var. *pallidus* which is green with short segment
flowers. Rupp suggests that it could be a hybrid
between *A. caudatus* and *A. exsertus*.

The Mosquito Orchid, *A. exsertus* (Plate 64)
occurs in all States except Western Australia and
the Northern Territory. It flowers in April to June,
and is very common and abundant in open forest
areas or grasslands in sheltered positions. Pixie
Caps, *A. fornicatus*, occur from the Clyde River in
New South Wales to Queensland (var. *sinclairii* is
found only in New Zealand). A plant resembling
A. exsertus, it is more robust, up to 30cm tall with a
heart-shaped leaf on a stalk above the ground. It
has up to ten variable reddish to green flowers.
Sepals and petals are narrow. The labellum is not
clawed, but oblong, lance-shaped with two raised
callus ridges on the lamina. The column is very
short for this genus. It is a reasonably common
species.

Var. *sinclairii* of New Zealand is a much smaller
form in all respects and the labellum is ovate,
extending to a fine point. *A. ledwardii* of southern
Queensland flowers in June and closely resembles
A. fornicatus. But it is only 5cm high with up to six
flowers, and somewhat depressed to the stem. The
dorsal sepal is deeply hooded and the labellum
almost lacking in calli. The Gnat Orchid, *A.
reniformis*, is found in all States except the Northern
Territory, also in New Zealand. The kidney-shaped
leaf distinguishes this taxon from the other species
in Australia. The New Zealand variety is *oblongus*,
named from the more oblong leaf. A wholly green
form of the type is known. The Western Australian
A. huegelii is a very rare plant, similar to *A. exsertus*,
but differing in its smaller flowers and its labellum,
which is broader at its base, tapering to an acute
apex.

Plate 63

Acianthus caudatus   DEAD-HORSE or MAYFLY ORCHID

New South Wales, Victoria, Tasmania, and South Australia. Flowers in the spring.

Plate 64

Acianthus exsertus   MOSQUITO OR GNAT ORCHID

All States except Northern Territory. Underside of leaf red. Autumn to winter.

Townsonia
CREEPING FOREST ORCHID

THE GENERIC NAME honours W. Townson, a New Zealand botanical collector.

T. viridis was originally described by Hooker from Tasmanian material under the name of *Acianthus viridis*. In 1906 Cheeseman of New Zealand described a plant under the name of *Townsonia deflexa*. During 1911 Schlechter removed *A. viridis* from *Acianthus* and placed it into Cheeseman's genus *Townsonia*. The Rev. Rupp[34] suggested in 1933 that the two were cospecific, with *T. viridis* being the oldest name.

Nothing is known of the pollination, but I would suggest that it could be self-pollinating owing to the anther position which overhangs the prominent stigma. As the pollinia becomes granulate it would drop directly on to the stigma.

Townsonia is closely allied to *Adenochilus* and *Caladenia*, differing from *Caladenia* in its creeping root stock, and from *Adenochilus* by a smooth undivided labellum. It also has small tubers.

T. viridis is found in a very restricted habitat in Tasmania only, growing in moss at the base of myrtle trees or in the moss on rotted logs. This species requires abundant moisture and occurs in beech forests at fairly high altitude.

T. viridis (Plate 65) occurs in Tasmania and New Zealand only, and flowers in December to January. Slender, up to 15cm tall. It has a flowering stem with no leaf, but a large floral bract below the flowers. The leaf rises from a separate stem at intervals along the rhizome. Flowers are small, one or two in number, and pale green to reddish, deflexed. The upper sepal is hooded and incurved, the lower sepals narrow, protruding straight under the hood or deflexed. The labellum is clawed and undivided, the lamina smooth, with no calli.

Liparis
THE TWAYBLADES

LIPARIS COMES FROM the Greek and means 'fat or greasy', alluding to the thick, fleshy, shiny leaves. It is a large genus of about 260 species, with its greatest concentration in the tropical areas of Asia and Oceania. Australia has approximately ten species. The Australian forms are divided into three groups, terrestrial, lithophytic and epiphytic.

The method of pollination in the terrestrial forms is open to discussion; some authorities[35] suggest self-pollination, but the Australian species set only few capsules to the flowers produced, suggesting cross-pollination by insects. So far no literature has been written on the pollination of our Australian species.

The Yellow Rock Orchid, *L. reflexa* (Plate 66) of New South Wales and Queensland, flowers in autumn and forms dense matts on rock-faces and ledges, rarely on trees. Flowers number up to thirty and have a very unpleasant smell. *L. nugentae,* found only in Queensland, is a larger version of *L. reflexa*, the plant being up to 30cm high. The flowers numbering about twenty, often less, are 2cm in diameter and a similar colour to *L. reflexa*. Also similar to *L. reflexa* is *L. cuneilabris*, differing in its wedge-shaped labellum and yellow flowers. The Fairy Tree Orchid, *L. coelogynoides* of New South Wales and Queensland, is the daintiest of the Australian species. An epiphyte, it forms dense mats on trees. The drooping raceme of pale green flowers, turns orange with age and has a hatchet-shaped labellum. Two terrestrial species are the deep reddish-flowered *L. simmondsii* of New South Wales and Queensland which has largish ribbed leaves up to 20cm and grows on the swamp margins. The very rare *L. habenaria* of New South Wales and Queensland has yellow flowers.

Plate 65

Townsonia viridis CREEPING FOREST ORCHID

Restricted to the myrtle forests in high rainfall areas of Tasmania, also New Zealand. Summer.

Photo: P. A. Palmer

Plate 66

Liparis reflexa YELLOW ROCK ORCHID

This autumn flowering orchid often grows in extensive colonies on cliff faces, rocks etc. in coastal districts of New South Wales and southern Queensland.

Corybas HELMET ORCHIDS

THE GENUS CORYBAS derives its name from one of the dancing priests of Phrygia. It consists of about fifty species extending from the Himalayan foothills east to the Philippines, and south-east through Malaysia to New Guinea, Australia, and New Zealand. It is also recorded from Polynesia. Nine species are found in Australia with two of these extending to New Zealand. Six other species of Helmets are found in New Zealand with one variety. In New Zealand *Corybas* are known as Spider Orchids, alluding to the much extended sepals and petals in the section *Nematoceras* which occurs there. How very inappropriate the name is in Australia, where we find all species have very minute sepals and petals.

The general habitat for *Corybas* is damp sheltered mossy or leaf-mould areas under low bushes or in open forest under bracken fern.

The pollination of *Corybas aconitiflorus* was discussed by Mrs Coleman[36] in 1931. In *C. aconitiflorus* it is the labellum which attracts the insect and also guides it to the pollinia. The structure of the column prevents this species being self-pollinating. At the base of the column a fleshy protuberance is filled with nectar. As the insect (which is not identified) probes the liquid it contacts the rostellum and viscid disc with its head, so that the pollinia become stuck to the insect's head as it leaves the flower.

The Veined Helmet Orchid, *C. dilatatus*[37] (Plate 67) occurs in all States except Queensland and the Northern Territory (July to September.) *C. fimbriatus*, of Tasmania, Victoria, New South Wales, and Queensland, which flowers in April to July. The flower is large, up to 3cm long, purplish with a white boss in the centre of the labellum; the lamina is large and the margins have a long fringe. The Frosty Helmet, *C. pruinosa*, occurs in New South Wales only. It resembles *C. fimbriatus*, but is usually smaller. Purple Helmet, *C. diemenicus*, occurs in Tasmania, Victoria, South Australia, and New South Wales, flowering in April to August (variable in different States). Similar to *C. fimbriatus* in leaf, which is kidney-shaped, with a smallish flower, the dorsal sepal is slate-coloured, the labellum grey to prune-coloured, the margins incurving and minutely toothed, the centre boss of the labellum white.

The Tailed Helmet, *C. undulatus*, is found in New South Wales and Queensland, and flowers in May to June. It is a small plant with a flower about 15cm in diameter and with a heart-shaped leaf. The flower is dark purplish-red with translucent patches. The prominently-veined segments and the labellum are produced into a tail. Abell's Helmet, *C. abellianus*, occurs in Queensland only, and flower is dark purplish-red with translucent patches. described from northern Queensland; its closest relation in Australia is possibly *C. aconitiflorus*, from which it differs in that the leaf is white-veined, and the labellum more open at its dilated apex, the centre being white. The Spurred Helmet, *C. aconitiflorus* (Plate 68) occurs in all States except Western Australia and the Northern Territory. It extends into New Zealand, and flowers in April to July. It is a common plant in some areas. The Small Helmet, *C. unguiculatus*, occurs in Tasmania, Victoria, South Australia, New South Wales and New Zealand (June to July). Together with the allied *C. fordhamii* it has a heart-shaped leaf, the small erect reddish-purple flower being on a long stem. The labellum is tubular with the gland inside on the lamina. Eastern Victorian occurrence of this orchid was noted in 1969.

Plate 67

Corybas dilatatus   VEINED HELMET ORCHID

Favouring cool mountain gullies and shaded coastal areas in Tasmania, Victoria, South Australia, and of rare occurrence in Western Australia. Late winter.

Plate 68

Corybas aconitflorus   SPURRED HELMET ORCHID

Occurring in eastern Victoria, Tasmania, New South Wales, and Queensland, also New Zealand. Autumn to winter flowering.

Cryptostylis TONGUE ORCHID

CRYPTOSTYLIS IS derived from *kryptos*, meaning hidden, and *stylos*, a column. This alludes to the column's concealment by the exceptionally large labellum, which encloses the very small column in its basal folds.

The flowers in this genus have the labellum dorsally located (at the top). The sepals and petals are all very narrow and inconspicuous. It is a genus of about twenty species, five occurring in Australia with the balance of the species in Formosa, the Philippines, Malaya, New Guinea, New Caledonia and Ceylon. A prominent stalked leaf is found in four of the Australian species.

Of the five Australian taxa, four are endemic. *C. ovata* occurs in Western Australia only. *C. subulata* is resident in Tasmania, Southern Australia, Victoria, New South Wales, and Queensland, extending then into the tropics. *C. leptochila* occurs in Victoria and New South Wales; *C. erecta* in eastern Victoria, New South Wales and Queensland, and the very rare *C. hunteriana* occurs sparsely in eastern Victoria and New South Wales.

The pollination of this genus aroused great interest when it was first described by Mrs Edith Coleman in 1927. If it had not been described by so eminent a person, it might well have been doubted by the scientific world. The method used is the pseudo-copulation of the plant by the male ichneumon wasp *(Lissopimpla semipunctata)*. The male wasp is usually hatched before the female of the species and apparently becomes attracted to the *Cryptostylis* by a scent similar to the female wasp. The wasp alights upside-down on the labellum and proceeds to back into the column, attempting to mate with the flower. During this act the pollinia of the orchid are deposited on the thorax of the insect. Wasps will often fight over the possession of a particular flower. It has been stated that the flower lures the male wasp more effectively than the female wasp. (See frontispiece for illustration of the pollination of *C. leptochila*). This could be due to the similarity in shape and markings of the labellum in this genus to the female body of the pollinating wasp but is generally assumed to be a scent attraction. Pollination by pseudo-copulation is now quite well known in other orchid genera of the world.

The Large Tongue Orchid, *C. subulata* (Plate 69) was described in 1806 by Labillardière, the French botanist, who named the plant *Malaxis subulata*. But in 1810 Brown named the species *Cryptostylis longifolia*, and it was left to Reichenbach to correct the identity of the species. The plants prefer a damp heathy habitat, the margins of swamps in eucalyptus forest or dry sandstone conditions around soaks or damp rocky crevices.

The Broad Leaf Tongue Orchid, *C. ovata*, resembles *C. subulata* in flower-shape and colour, but has a large ovate leaf. The Small Tongue Orchid *C. leptochila*, is more of a highland species in New South Wales, whereas in Victoria it is found in the south-west of the State, almost on the coast. It has a narrow and smallish leaf; the labellum is narrow and much rolled to the top of the lamina, the underside having a series of very conspicuous glands either side of the midrib. The Tartan Tongue Orchid, *C. erecta*, is an attractive species. The labellum is formed into a conspicuously veined bonnet. The Leafless Tongue Orchid, *C. hunteriana* (Plate 70) is regarded as a saprophyte. Usually lacking a leaf, it resembles *C. leptochila*, but the labellum is broader and densely furred. This species, discovered at Marlo, eastern Victoria, extends to New South Wales. It is very rare.

Plate 69

Cryptostylis subulata   LARGE TONGUE ORCHID

Present in all eastern States and South Australia, generally in marshy areas. Summer.

Plate 70

Cryptostylis hunteriana   LEAFLESS TONGUE ORCHID

A rare plant of eastern Victoria and southern New South Wales. Summer flowering. (See frontispiece.)

AUSTRALIAN SAPROPHYTIC ORCHIDS

IN AUSTRALIA there is a particular group of usually leafless, mostly summer flowering plants that rely on a symbiotic association with various ground fungi. This association has been described in the Introduction.

Saprophytes can be found in many genera of orchids; they may contain the whole genus or only one or two species of a normally green-leafed genus.

Australia has approximately twenty-one species that are classified as saprophytes. New Zealand can add three more to this total and these will also be included in this section.

Some plants almost fall into the saprophyte group, for example *Spiculaea huntiana*, a plant regarded as leafless, but as with *Burnettia cuneata* and other of the *Spiculaea* genus it may produce a leaf when not in flower. Further observations are required.

Most saprophytes are found growing in soil well supplied with rich leaf-mould. In New South Wales they are found under *Eucalypt* trees in open forest, under *Melaleuca* trees by the beach, and in rain-forests.

Saprophytes are at times referred to by the layman as parasites, but this is not the case, as the saprophyte lives only on dead and decaying matter, whereas the parasite attaches itself to and derives nourishment from living organisms.

Most of the Australian saprophytes apparently can live without the fungus for short periods. This occurs when they are removed by people in an attempt to cultivate these very attractive flowers. Usually the plant will survive for a season before dying. It has been living off its stored nutriment in the very thick and fleshy root system common to all members of this group.

Many of the species are cross-pollinating. Rogers[38] describes as follows the pollinating method used in *Dipodium punctatum*. The splitting of the anthers occurs just as the floral segments begin to expand; the pollinia will at this time still be sticky on their lower surfaces. A little later, just as the flower is fully open, the anther cap becomes very hard and the pollen masses are liberated, each dropping upon the pointed caudicle adhering to its underside. The pollinia are guided into position by a small band of tissue connected from the rostellum to the top of the column. The pollen masses eventually become very hard, but the sticky fluid that attaches them to the caudicles never really hardens, allowing the pollinia to be easily removed from the caudicles. The labellum in *Dipodium* is very rigid and sessile to the column, affording a good landing place to any visiting insect. The agent (identity not known) is guided to the rostellum and its viscid disc by the small side-lobes of the labellum, and is stopped going too far into the flower by the pad of bristle-hair on the labellum. The pollinia are removed only on the insect's withdrawal from the flower. As soon as the pollinia are removed the caudicles curl, taking the pollen masses and placing them in a forward position ideally suited to contact directly the stigma cavity of the next *Dipodium* visited.

Of the self-pollinating species Hatch[39] describes the method of pollination in two of the New Zealand *Gastrodia* species, *G. cunninghamii* and *G. minor*.

The column in these two species, which are possibly derived from the ancestral stock of the Papuan species *G. papuana*, has a series of concertina-like rolls on the back of the column, which is in an upright position before the anthers ripen. But as ripening occurs the rolls expand, curving the column and taking the ripe pollinia with it, pressing this hard on to the stigma which lies on

Plate 71

Eulophia venosa

A plant of coastal areas of northern Queensland and
Northern Territory. October–November flowering.

Photo: F. M. Coate

Plate 72

Galeola cassythoides CLIMBING ORCHID

Often exceeding 5m in height and climbing living or
dead trees, this orchid extends north from central New
South Wales to Queensland. September–November.

Photo. W. Upton

the basal section of the column. In the Australian species this procedure certainly does not occur. *G. queenslandica*, one Australian species, is self-pollinating, but the pollination is virtually unknown as its life cycle occurs in only a few days. *G. sesamoides*, the other Australian species, appears to be insect-pollinated, as in fifty-eight flowers in the author's herbarium only eleven showed fertilised capsules.

Cryptostylis hunteriana is pollinated as described in that genus, as is *Prasophyllum flavum* and *Calochilus herbaceus*.

Possibly the commonest plant in this section is the Potato Orchid *Gastrodia sesamoides* (Plate 73). It occurs in all States and New Zealand, and is one of a small genus of about twelve species, chiefly Indo-Malayan. The Australian species is slender, rising from a potato-like tuber. The cinnamon-brown flowers with a whitish apex form a terminal raceme. The segments are united into a bell-shape. The basal portion is swollen (the generic name means pot-bellied, alluding to this feature).

G. cunninghamii of New Zealand differs from *G. sesamoides* in the column being shorter than the labellum and the back of the column having concertina-like rolls. *G. minor* of New Zealand has straight column wings and a broad base to the labellum, usually smaller than the other two species. A recently described species from North Queensland is *G. queenslandica*. This species does not resemble *G. sesamoides* at all, being very short, the ovaries not swollen, and it has two tuberculate calli on the base of the labellum. This species is self-pollinating. Very closely allied to *Gastrodia* is the Northern Territory plant *Didymoplexis pallens*. This plant differs from *Gastrodia* in that the petals are disunited from the lateral sepals and the stigma is near the top of the column. The species is found from India to Java, Sumatra through to Australia. Stems are up

to 12cm high with a few pale brownish-olive to pinkish flowers, 1 to 1.4cm long. The upper sepals are united to petals, the lateral sepals united, but only to the petals at the base. The labellum is whitish-yellow. Plants of this species are hard to see in flower, but the long stalked fruit can more easily be detected.

Dipodium, from the Greek *dis*, meaning double and *podium*, a little foot, refers to the two false caudicles of this genus. It is a small genus of about ten species falling into three sections, epiphytes, green-leaved terrestrial herbs, and terrestrial saprophytes. The latter two sections are represented in Australia. The epiphyte section is only found in the Malaysian area[45]. The commonest Australian species is the Hyacinth Orchid. *D. punctatum* (Plate 74), which occurs in all States apart from Western Australia, flowering in January to April. The variety *stenocheilum* is a northern form of the species from northern New South Wales to Cape York in Queensland. The plant is more slender than the type, and the flowers are smaller, unspotted, and with segments narrower and blunter. The base of the labellum is narrow, distinctly sac-shaped. A white variety, *album*, is also known from the south coast of Queensland. *D. hamiltonianum* closely resembles *D. punctatum*, but is larger, its bright yellow or greenish-yellow flowers spotted and streaked with red. Chiefly found on the tablelands and western slopes of New South Wales, it also extends to Queensland. The third species is *D. ensifolium* of tropical Queensland, which has sword-shaped leaves and is found in clumps. The flower stem is similar in shape and colour to *D. punctatum*. This latter species *(D. ensifolium)* cultivates well in cymbidium mixture enriched with cow manure.

Two unique Australian plants are the subterranean orchids *Rhizanthella* and *Cryptanthemis*. Both genera belong to the subtribe *Rhizanthellinae*, the

Plate 73

Gastrodia sesamoides   POTATO ORCHID

Recorded from all Australian States except Northern
Territory, also New Zealand. Late spring to summer.

Photo: B. A. Fuhrer

Plate 74

Dipodium punctatum   HYACINTH ORCHID

Though usually spotted pink, flowers can be unspotted,
brown, yellow, or white. All States except Western
Australia. Generally summer flowering.

former *R. gardneri* (Plate 76), was ploughed out in 1928 by John Trott at Carrigin, Western Australia, in a burnt over area. The orchid was found in close association to the rotting roots of a *Melaleuca* sp. the ground surrounding both the stump and orchid was heavily impregnated with an *Aspergillus* fungus.

Cryptanthemis slateri (Plate 75), was also accidentally collected in 1931 by E. Slater, growing in conjunction with another saprophyte *Dipodium punctatum* in a dry watercourse near a disused Alum quarry at Bulahdelah, New South Wales. Rupp states that the flower head matures below the ground and is pollinated possibly by a burrowing insect, then they rise to the surface of the soil to disperse seed. This species also occurs in South Queensland.

The species *Epipogum roseum* occurs in Africa, India, Malaysia, Queensland, and New South Wales. It is a remarkable plant with a weak drooping (at times erect) raceme of many white or pink flowers. A small spur points from the base of the labellum to below the ovary.

The life cycle of this plant is completed in only a few days[40] from breaking ground till the dispersal of the seed. The Great Climbing Orchid, *Galeola foliata*, occurs in New South Wales and Queensland (November to January). It is possibly the tallest orchid in the world, growing up to 12m high. The flowers are numerous and bright golden-yellow. The labellum is white with pink markings with a glandular ridge along its middle. The pollen is granular. *G. cassythoides* (Plate 72) is found from the Illawarra area of New South Wales to Queensland (September to November). It is a smaller plant, growing up to 6m. It is freely branching. The genus *Eulophia* has four species, two of which are saprophytes. None are common, and all are similar. *E. venosa* (Plate 71) has thin deciduous grass-like leaves, as has *E. agrostophylla*. *E. fitzalanii* is leafless and has a finely-veined labellum. The mid-lobe is as long as it is broad. *E. zollingeri* has flowers 2.5cm wide, of a dark purplish colour. The New Zealand *Yoania australis* resembles *Burnettia* (Plate 55) in habit, but has a rhizome and grows in deep litter beneath taraire trees. The genus *Aphyllorchis*[41] has two rare species, *A. anomala* and *A. queenslandica*, both brittle herbs from Queensland. Finally the genus *Genesplesium* has one species *G. baueri* which closely resembles a fleshy *Prasophyllum*, but it has a tubular-shaped labellum and very long divergent lateral sepals. It is very succulent and brittle in habit. Found growing only in sandstone areas of New South Wales.

Plate 75

Cryptanthemis slateri   EASTERN UNDERGROUND ORCHID
This rare underground orchid from the Bulahdelah area
of New South Wales was drawn from material published
by H. M. Rupp in 1933–4–5. (Proceedings of Linnean
Society of New South Wales.) Also found in south
Queensland.   *Drawing: M. I. Barnes*

Plate 76

Rhizanthella gardneri
                    WESTERN AUSTRALIAN UNDERGROUND ORCHID
As rare as *Cryptanthemis slateri* and recorded a few times
only from Western Australia. Drawing based on C. A.
Gardner's papers and other unpublished illustrations.
                                     Drawing: M. I. Barnes

Dendrobium AND ALLIED GENERA

THE GENERIC NAME comes from the Greek *dendron*, a tree and *bios*, life, as the majority of species are found growing on trees as epiphytes. A few, however, are lithophytes (growing on rocks).

This is a very large genus of approximately sixty Australian species, confined mainly to the tropical and subtropical areas of Australia but extending southwards with two species, *D. speciosum* and *D. striolatum*, occurring in Victoria, and *D. striolatum* extending through the Bass Strait islands to Tasmania, where it is found at altitudes up to 3,000ft, and is at times snowbound. The southernmost record of a *Dendrobium* must go to New Zealand's only species *D. cunninghamii*, and its Stewart Island occurrence.

The earliest description of a *Dendrobium* was of *D. linguiforme* by Swartz[7] in 1800.

In a book of this brevity only few of the vast number of forms can be considered. No comprehensive work on this genus has been undertaken since Rupp and Hunt's[46] work in 1948. Much has been added to our knowledge of this genus since its publication.

It is intended to divide the discussion into four sections, 1 Tropical species, 2 Temperate and subtropical species, 3 Plants with fleshy leaves, and 4 Allied genera.

Pollination of *Dendrobium* is by insect agent. R. Fitzgerald, in *Australian Orchids,* attributes the pollination of *D. speciosum* to the *Dendrobium* beetle *Stethopachys formosa*, stating that although this beetle attacks the plant he has observed that the plant produces capsules only after attendance by the beetle.

Quite a number of possible natural hybrids have been recorded in this genus. These are described below:

D. x *superbiens* (back jacket) resembles *D. discolor* in general plant form; the flowers are similar to the *D. bigibbum* complex, but elongated in segments which are often flecked on the margins with white. *D. nindii* x *D. discolor* resembles *D. discolor*, but the stems are flattened, as in *D. nindii;* the flowers are at first flushed with yellow, later changing (except at the base of the labellum) to a dusky pale mauve. It resembles both parents of the typical antelope orchid-type flower (section *Ceratobium*). *D.* x *gracillimum* and *D.* x *delicatum:* (in this species I include the controversial *D. kestevenii* as it has the same parents as *D. delicatum*, so must be placed under the oldest name). These plants have been so well discussed in recent years that reference to the Australian Native Orchid Society Journal *The Orchidian* is recommended if further particulars are required. *D.* x *suffusum*, a recent addition, resembles *D. gracilicaule* in plant habit; the flowers resemble *D. kingianum* in shape and take their colour from this species, being white inside, with a spotted exterior surface suffused with mauve-pink. The labellum is also variously marked with mauve-pink.

1a *Tropical hardwoods* (Phalaenanthe)

These are referred to as the *D. bigibbum* group. All have large flowers of great horticultural value and the better forms have been extensively used by the orchid hydridiser, with spectacular results. Plants usually have slender stems up to 1m high, with numerous large flowers to 6cm, in long arching racemes up to 45cm long rising from the apex of the plant. The colour varies from white through all shades of pink to magenta. The type form of *D. bigibbum* has a small bunch of white glands on the labellum; var. *superbum* is much superior in flower size and colour and has large mothlike flowers. A sub-variety of *superbum* is *compactum*, which is similar to the variety but with short stems,

Plate 77

Dendrobium monophyllum LILY-OF-THE-VALLEY ORCHID

Often forming patches on trees and rocks. Each bulb usually has one leaf. Queensland and New South Wales. Spring.

Plate 78

Dendrobium canaliculatum ANTELOPE ORCHID

Confined to Queensland only. August–October flowering. Also called Tea-Tree Orchid in north Queensland.

rarely above 10cm in height. This complex of orchids is found in Northern Queensland, flowering mainly in the autumn and early winter, occurring mainly in dry areas on trees and rocks.

D. dicuphum (Plate 81) is found in the Northern Territory and the north-west of Western Australia. Though similar to *D. bigibbum* it has smaller flowers. In its typical form it is white with a red suffusion at the base of the labellum, the middle lobe of the labellum is much longer than that of *D. bigibbum*. A variety *grandiflorum* is named, having similar flowers, half as large again as the type.

1b *Tropical Antelope Orchids* (Ceratobium)

D. discolor (syn. *undulatum*) is one of the largest *Dendrobiums* in Australia, with a stem at times up to almost 3 metres long. This species occurs from near Rockhampton in North Queensland to New Guinea and adjacent islands, and flowers chiefly in the late winter to spring. Flowers are typically brownish with yellow suffusions, the labellum lightly veined with purple. The petals and sepals are variously twisted. A clear lemon-yellow forma *broomfieldii* is known. Similar to this species is *D. nindii*, a magnificent plant close to *D. discolor* in plant form. The flower-segments are twisted and white. The large roundish three-lobed labellum is marked and veined with purple. A plant resembling a large *D. aemulum* but with flowers resembling *D. discolor* var. *fuscum*, though not all twisted, is *D. wilkianum*. The flowers are dull brownish, the labellum a yellowish-green with numerous transverse red lines across the lateral lobes. *D. johannis* has flowers up to 4.5cm across. The segments of approximately equal length, are all narrow, brown, with yellowish and greenish tints. The labellum is three-lobed. The lateral lobes are deep red and veined, the mid-lobe is yellow. A plant with a greenish flower only 2cm in diameter (very localised

around Cairns) is *D. bifalce*. *D. smillieae* is conspicuous with its dense short raceme of waxy white, yellow, pink, red and green-marked flowers, being common in the lowlands of Northern Queensland. *D. canaliculatum* (Plate 78) has an onion-like pseudo-bulb topped with thick roundish leaves; its long spray of twisted flowers is very attractive. This plant occurs on paperbark trees (*Melaleuca* sp.) in the drier lowlands of Northern Queensland and the Northern Territory, and flowers in the spring. Many other species occur in this group.

2. *Species occurring in temperate and subtropical regions.*

This section contains many plants, the commonest being *D. speciosum*, the Rock Lily or King Orchid (Plate 80). It occurs from Northern Queensland south to eastern Victoria and is known in four varieties and two forms. It flowers in the early spring. *D. ruppianum* (syn. *fusiforme*) is another fairly common species, the cigar-shaped stems, topped with four or five dark green leaves, are very attractive with the long dense racemes of narrow segmented creamy white flowers. The labellum is mauve-spotted. Spring flowering, it occurs in Northern Queensland. The flowers of the 'Beech Orchid' of New South Wales are highly perfumed in sunshine. It grows high in the beech forests (*Nothofagus moorei*) and has stems to 20cm long with a terminal raceme of glistening white flowers with a purple marked labellum, the apex of which is upturned into a sharp beak. Though spring flowering, plants are often snowbound in winter. *D. kingianum* is a cosmopolitan lithophyte, normally a smallish plant with stems of 8 to 10cm, but forms with stems of 50cm are known. The flowers are very variable, ranging from pure white to a deep magenta-purple. It occurs in New South Wales and Queensland. The spring-flowering 'Orange and

Plate 79

Dendrobium striolatum
STREAKED ROCK ORCHID

Generally growing on rocks and ranging from eastern Tasmania through the Bass Strait islands and eastern Victoria to New South Wales. September–November flowering.

Plate 80

Dendrobium speciosum
KING ORCHID or ROCK LILY

The largest epiphytic orchid in Australia. Far-eastern Victoria, New South Wales and Queensland, with several known varieties. September–November.

Brown' *D. gracilicaule*, is a slender plant with tall thin stems to 70cm long. The small clusters of yellowish orange flowers marked and suffused with red-brown are very attractive. A variety *howeanum*, with rich creamy yellow to whitish flowers with a faint pink lip occurs on Lord Howe Island, but has once been recorded from near Gosford, New South Wales.

D. tetragonum, with its square pendulous stems, is a very variable New South Wales and Queensland species. It divides into two varieties, var. *tomentosum*, which has a 'felted' labellum, and var. *giganteum* with flower to 10cm across, greenish, marked with red. The labellum has three lobes. The type form has small spidery yellow, green, red and white flowers. The White Feather Orchid, *D. aemulum*, of New South Wales and Queensland, has slender stems topped with four or five oval leaves and is found in two distinct forms. A short stunted type grows on the ironbark eucalypts and a long-stemmed form occurs in rain-forest. A spray of feathery white flowers gives this species a very dainty appearance, and the flowers turn pink with age. The Lily of the Valley Orchid, *D. monophyllum* (Plate 77), flowers most of the year and occurs in New South Wales and Queensland. Similar to this species is *D. schneiderae*, found on the hoop pines of northern New South Wales and Queensland.

3. *The fleshy-leaved species*

These mainly grow from a creeping rhizome and form dense clumps on their hosts. They can be divided again into plants with flat leaves and plants with pencil-like leaves. *D. linguiforme* of New South Wales and Queensland has flat oval leaves resembling a thumbnail with a spray of feathery white, rarely cream, or yellow flowers. Var. *nugentii* from Northern Queensland has shorter segments

and a smaller labellum. *D. rigidum* resembles the above species, but is more pendulous in growth and the two or three yellowish, flushed pink flowers form at the base of the leaves. The labellum is purple-marked. *D. cucumerinum* has leaves like small gherkins, and flowers from the base of the leaf; it is cream to greenish, and the labellum is marked with purple on the three ridges. The 'pencil' types usually have long terete leaves, pointing either up or down. The common *D. teretifolium* has long bean-like pendulous leaves. In August it has sprays of creamy flowers. A large example of this species is a memorable sight.

Two other varieties are named—var. *fairfaxii* from the rain-forest areas in New South Wales, a form *aureum* which extends into Queensland; and the var. *fasciculatum* found in Northern Queensland, which is more robust in plant form. *D. beckleri* has erect pencil-like leaves and is more robust than the former plant. The flowers are white and marked with purple on the lip. *D. striolatum* (Plate 79) is a rock-dweller with short slender leaves. At times it has short branches. It flowers in the late spring with yellow red-streaked flowers. A species with long slender leaves is the rare *D. tenuissimum*, which occurs in brush-forests from the Barrington Tops area to southern Queensland. It flowers in the spring. The small purplish-green or green flowers have the white labellum marked with bright purple towards its apex. A very common plant is the Dagger Orchid, *D. pugioniforme*. It flowers in October to November and is found from the Queensland border in the north to Mt Dromedary in southern New South Wales. This species forms dense clumps on trees and may hang down some 180cm. The long dagger-shaped leaves often hide the shy, light-green, purple-marked flowers. This species prefers shade in its habitat.

Plate 81

Dendrobium dicuphum

Occurring in Northern Territory, and north-western Western Australia, this orchid is October–November flowering at Darwin.

Plate 82

Eria fitzalani

The largest of the Australian *Erias*, this species is confined to northern Queensland. Spring.

Allied genera

The genus *Cadetia*, closely allied to *Dendrobium*, has two species in Queensland, *C. taylori*, and *C. maideniana*. Both species have white flowers, and short stems with one apical leaf. *C. maideniana* differs from *C. taylori* in having the ovary densely covered with short thick hairs. Both plants grow in dense tufts.

The genus *Eria* contains six species in Australia four of which, *E. eriaeoides*, *E. irukandjiana*, *E. dischorensis* and *E. queenslandica*, have small insignificant flowers, while *E. fitzalani* (Plate 82) and *E. inornata* have larger more decorative flowers. All species are indigenous to Queensland and most have roundish to oblong pseudo-bulbs with strap-like leaves from the apex of the pseudo-bulb. The flowering period is from August through to December.

The genus *Diplocaulobium* has two species, *D. masonii*, endemic to Australia, and *D. glabrum*, found in Queensland and extending to New Guinea. *D. masonii* resembles *Dendrobium monophyllum* in habit while the flowers are similar to *Dendrobium tetragonum*. *D. glabrum* is similar in habit. The pseudo-bulbs are about 4cm by 1cm, topped with a single long ovate leaf. It has only one flower, short-lived, on a long stem, pale yellow, the slender segments transparent and about 3cm long. The mid-lobe of the three-lobed labellum is prominent.

The genus *Ephemerantha* contains only two species from Northern Queensland. It closely resembles *Diplocaulobium*, an epiphyte with a creeping rhizome. The pseudo-bulbs are widely spaced. The single apical leaf is on a short slender stalk. The fragile flowers (lasting less than a day) occur from the leaf base and the labellum is obscurely three-lobed with the mid-lobe divided.

This group cultivates well in general, requiring a cool glasshouse or bushhouse, with plenty of water in the growing season. Many are shy to flower.

Geodorum

THE GENERIC NAME of this genus means literally 'flower being near the earth', alluding to its peculiar habit of the decurving flower raceme, thus taking the labellum to a low position. This is possibly to assist the pollinating agent, for after fertilisation of the ovaries the flower raceme returns to the upright position for seed dispersal. It is not a large genus, containing about ten species, only one occurring in Australia. The remainder are chiefly Indo-Malayan. The Australian species occurs in New South Wales and Queensland and is apparently endemic. Bentham *(ibid)* suggests that *G. pictum* may be cospecific to the Indian species *G. dilatatum*.

The pollination of this species is probably carried out by insects.

G. pictum (Plate 83) is a largish plant with broad, prominently-ribbed leaves. The dense terminal raceme of pink flowers are about 2cm each in diameter, and have a broad rounded base to the labellum, which is somewhat constricted at its centre, and the apex is acutely pointed. The lamina is prominently veined.

Plate 83

Geodorum pictum   PAINTED ORCHID

Occurring in northern New South Wales, Queensland, and Northern Territory. Spring flowering.   *Photo: C. W. Harman*

Plate 84

Calanthe triplicata   SCRUB LILY

The flowers of this orchid bruise easily. Found in New South Wales, Queensland, and Northern Territory. Occasionally called the Christmas Orchid, because of its December flowering.

Photo: C. W. Harman

Calanthe SCRUB LILY

CALANTHE TRIPLICATA (Plate 84, syn. *C. tripli-cata*) is one of the more exotic of our ground orchids with large strongly-ribbed deep green leaves, up to 90cm high. The flower spike, which is topped with a cluster of fairly large white flowers, can be up to 150cm high. It is a large genus with up to 120 species, only one occurring in Australia. This species extends from New South Wales to Queensland and through Malaysia to India and China. Very little is known about its pollination; some of the exotic species are self-pollinating, but it is felt that *C. triplicata* is insect-pollinated, as it rarely sets seed in cultivation unless hand-pollinated.

It has flowers that are easily bruised, turning black if they are touched. If the pollinia are removed the pure white florets turn yellowish.

'Scrub Lily' is a misnomer, as it certainly is not a lily and it is not confined to scrub locations. It is found most frequently alongside creeks in fairly heavy shade, where it grows occasionally in large numbers. In cultivation this species requires a reasonable amount of shade and an abundance of water, particularly in the summer months. It also relishes a liberal amount of manure in the compost, as it naturally grows in deep, rich leaf-mould, where its roots form a dense mat and may travel for several metres.

Calanthe is from the Greek and literally means 'pretty flower'.

Phaius SWAMP LILY

THE GENUS PHAIUS, although not a large one, is widely distributed, being found in New South Wales and Queensland, extending to New Guinea, the Pacific Islands, Indo-Malaya and tropical Asia.

P. tankervilliae (Plate 86) is by far the oldest-known of the Australian forms, having been introduced to England in 1778 by Sir Joseph Banks. It was originally described by Banks from plants collected in China as *Limodorum tankervilliae*. It is often listed in collections as *P. grandifolius*.

It is a large plant to 210cm high with large ribbed leaves to 120cm and unusually large 10cm-diameter flowers. The habitat is in swampy or boggy places, with its roots often in water most of the year. It grows well in cultivation and should be kept wet during all hot weather, with a reasonable amount of shade. Close to this species, is *P. australis*. W. H. Nicholls in a paper[42] on the Australian species upholds *P. australis* as a valid species, although it is often synonymised under *P. tankervilliae*. *P. australis* differs in its darker leaves; its flowers are deep red-brown inside with yellow- to chocolate-brown veins near the segment tips. The back of the flower is similarly coloured but lighter, and the labellum is not as tightly tubed. The variety *bernaysii*, found only in southern Queensland, is wholly sulphur-yellow. This form is often included under the exotic *P. blumei*.

Our third and last species is *P. pictus*, a recent addition which has smaller flowers of a deep buttercup yellow. The four-angled stems can be 60cm high. This species grows above 2,000ft in the Bellender Ker Range of Queensland.

Plate 85

Vanda tricolor

Confined in Australia to Northern Territory, where very rare, and extending to Indonesia. Spring. *Vanda whitaena* of northern Queensland on front cover is also a rare plant.

Plate 86

Phaius tankervilliae   SWAMP LILY

In coastal and tableland areas of northern New South Wales and Queensland extending to Asia. The largest of the Australian terrestrial orchids. Spring.

GENUS *Bulbophyllum* IN AUSTRALIA

BULBOPHYLLUM, WHICH MEANS literally a bulb-like leaf, is a good description of some members of this genus. It is now an extremely large genus which is divided into numerous sections. We cannot consider these in such a brief account as this, but give details only of the species commonly found in Australia.

This genus contains upwards of 1,000 species, some of which are closely allied and at times very difficult to separate from the genus *Dendrobium*. In fact some of our small northern Queensland *Dendrobiums* (*D. lichenastrum* complex) were originally described as *Bulbophyllums*. Australia has about 21 species. A. W. Dockrill, who has reviewed the genus in Australia, has divided the species into twelve sections, and due to the difficulty of assigning the Australian plants to the right section he has mainly used numbers for each section.

The pollination of this genus has had little written about it; one account of an exotic species mentions the attraction of blowflies to the foul-smelling flowers. It is thought all Australian species are insect-pollinated, as they do not readily set pods.

Bulbophyllum is essentially a rain-forest-dweller and in Australia is chiefly found in the coastal rainforest areas from northern Queensland to the Clyde River in New South Wales. Some species, such as *B. exiguum* and *B. crassulifolium*, can also be found in moss on the shady side of rocks in dry sandstone areas.

The genus responds well to cultivation and many grow easily in a bushhouse in broken shade in moist conditions. The plants are usually tied to paperbark or slabs of tree-fern fibre or fern peat, and hung from the roof. They require plenty of water and respond well to artificial manure solution.

The flowers of the Australian species are generally small, but often what is lacking in size is made up for by quantity. A rock covered with hundreds of dainty cream flowers of *B. exiguum* is a sight to behold.

The largest of our species is *B. baileyi,* which flowers in late spring to summer and occurs only in northern Queensland. The bulbs are crowded with ovate to oblong-ovate leaves on an oblong-shaped pseudo-bulb. The solitary flowers are white or cream, spotted with purple. A smaller species is *B. nematopodum,* whose leaves are long and lance-shaped. The solitary greenish flower is at times almost translucent, the segments veined, and finely spotted with green. Common in some areas is the fleshy-leaved group. The Wheat Leaf Orchid, *B. crassulifolium* (Plate 88), is one of the most southern-growing species, forming dense masses on rocks and trees. The flowers, very small at the base of the leaves, are cream, with tips of the segments darker. Similar to this species is *B. gadgarrense,* differing in the longer leaves and the thin, almost filiform segments to the flower, which are tipped with dark orange. *B. aurantiacum* is common in Queensland and northern New South Wales and resembles *B. crassulifolium*, but has wide leaves and flower-segments tipped with bright orange. A terete leaf-form of this species is found, but it differs very little in its flower. *B. exiguum*, has small, crinkled or at times round pseudo-bulbs, topped with a small dark green oblong leaf. The flower is in a short raceme of up to four cream or white blooms. It is common in New South Wales, but not quite so common in Queensland. It flowers in the autumn. The only plant in Australia with a dense capitate (grouped together at the top of the stem) flower stem is *B. evasum*, which has a brittle creeping rhizome. The pseudo-bulbs are somewhat flattened and widely spaced on the rhizome. It occurs only in Queensland. A pair of rather rare

Plate 87

Bulbophyllum macphersonii

Confined to Queensland, usually at elevations exceeding 1,500ft. Chiefly autumn flowering.

Plate 88

Bulbophyllum crassulifolium

WHEAT-LEAVED BULBOPHYLLUM

Occurring on trees and rocks in New South Wales and southern Queensland. Chiefly spring flowering.

miniatures are *B. minutissimum* and *B. globuliforme*. The former is mainly confined to *Ficus* trees or rocks, and occurs as far south as Milton, New South Wales. It has very small pseudo-bulbs, about 1.5mm across. A minute lance-shaped leaf appears at times, and forms the centre of the bulb. The solitary flower is purple, very hairy, with translucent patches. The latter species has a smaller pseudo-bulb, and the flower is white. This plant appears to be confined to the hoop pine forests of southern Queensland and northern New South Wales. *B. macphersonii* (Plate 87) and its variety *spathulatum* is the only Australian species without twisted ovaries. The leaf is wheatlike, very dense. The solitary flower is held well above the leaves. Var. *spathulatum* has a labellum shaped like an ice cream spoon. It flowers in the early summer to early winter. *B. weinthalii* is confined to the hoop pines of the northern New South Wales area and has greenish, spotted purple flowers as large as its pseudo-bulbs.

B. elisae (Plate 89) is a very distinct plant, as it has light green, warted pseudo-bulbs topped by an oblong darker green leaf. The long raceme of green flowers has extremely long lateral sepals.

One section of the genus often found in orchid collections is *Chirrhopetalum*. Only one species of this section is found in Australia, *B. clavigerum*. This plant has a creeping rhizome, covered with long hairs. The conical, deeply-ribbed and furrowed pseudo-bulb is about 5cm high, topped with a thick oblong short-stalked leaf. There are up to seven dull yellow, purple-spotted flowers which form a half-circle on the end of the long slender flower-stem. The lateral sepals are conspicuous, and are joined for about a quarter of their length to the tip. The labellum is tongue-shaped, much curved, mobile on a short claw. This rare plant is found on Cape York Peninsula, northern Queensland.

Oberonia

THE GENERIC NAME is derived from Oberon, the 'King of the Fairies'. This is a large genus of plants with at least 150 species, distributed from East Africa to Samoa. Only three species are found in Australia, all of them occurring in Queensland with two extending into New South Wales. *O. muellerana* was determined by Baron von Mueller in 1865 from plants collected by the Brisbane River as *O. iridifolia*. Mueller was not sure of his decision, but it remained unchallenged until 1907, when R. Schlechter published the now-accepted specific name. The Rev. Rupp cleared up this tangle in 1948. *O. palmicola* was also erroneously named *O. titania*, a name that was used for many years. But the Australian plant has been found not to be cospecific with the tropical species of this name.

The third species, *O. attenuata*, was recently described (1960) by A. W. Dockrill from plants collected in northern Queensland on the Mossman River. *O. attenuata* is a pendulous plant, with a few fibrous roots, the dark green leaves (up to seven) gradually tapering to an almost filiform point, the longest about 15cm. The raceme is as long as the longest leaves. The minute flowers are red-brown and appear in June. Smaller than this species is *O. palmicola* (Plate 90), with about five leaves, to 6cm long. The raceme is of minute brick-red flowers, arranged in whorls around the stem. It flowers in the spring. The third species, *O. muellerana* has flowers, slightly larger than in *O. palmicola*, which are yellowish green in a curved raceme. This genus cultivates fairly well if given a fair amount of shade in a damp cool glasshouse or bushhouse with at least one good watering each day. Nothing is known of the pollination of *Oberonia*.

Plate 89

Bulbophyllum elisae

Ranges from the Blue Mountains to south-eastern Queensland. Generally on rain forest trees at altitudes above 2,000ft. May to November flowering.

Plate 90

Oberonia palmicola

SOLDIER'S CREST ORCHID

The smallest of Australia's three *Oberonia* species. Northern New South Wales to central coastal Queensland. Spring to late summer.

Cymbidium

THE GENERIC NAME means 'boat shaped', an allusion to the labellum shape in some of the species. The *Cymbidium* genus consists of about fifty terrestrial and epiphytic plants and is distributed from Madagascar to Ceylon, India, Japan, through to Australia. Most of the species are very decorative; some are highly prized for cultivation and have been extensively used for hybridisation. Plants of the true species are now rare in general cultivation.

Cymbidium has three species and one variety in Australia. The only species that readily accepts cultivation is *C. madidum*, which thrives in standard commercial *Cymbidium* mixture. *C. suave* and *C. canaliculatum* are both very difficult to grow, persisting for a year or so but invariably dying. They are generally moderately large plants and usually grow in the hollows of trees, where their roots penetrate deep into the old rotten heart wood of the trees. Another favoured position is on *melaleuca* (paperbark trees) where the roots may run under the top layer of bark and travel all over the tree. Any root disturbances in the latter two species usually prove fatal. All three species are widespread and very variable in shape, colour, and size of plant. For instance the variability of *C. canaliculatum* can be illustrated by the number of forms that have been named in the past; these now have been rightly synonymed under the type in a recent review by A. W. Dockrill.[43]

The pollination of the Australian species is by insect. In *C. suave* the author has seen the flowers attended by a common native bee (not identified), but have not seen the actual pollination effected. Whatever the agent's method it is most effective, as two plants in cultivation at Kiama contain twenty-three fertilised capsules out of fifty-six flowers. The capsules take approximately three months to ripen and contain thousands of dustlike seeds.

C. canaliculatum (Plate 91) is an extremely variable species. Its distribution is from about the Hunter River in New South Wales north to Cape York, northern Queensland, and westwards to northern Western Australia. It is not normally a coastal plant, usually keeping to the drier inland areas. The stems have a swollen base (pseudobulb) with thick channelled leaves. The plant is an overall grey-green in colour. The flowers can be green, purple, brown, dull red or combinations of any of these colours, and can also be variously spotted. This plant grows into large clumps and with one or two racemes to a bulb many thousands of flowers can occur on a large plant. The flowering period is from September to December. *C. madidum* (Plate 92) is a tall plant to 25cm high resembling the typical *Cymbidium* in habit. The flower raceme is pendulous and attains a length of 65cm with up to eighty flowers, usually with a wider spread on the stem than in the other two species. The olive-green or golden-green labellum has a brown to blackish mark on the lamina. It occurs from about the Hunter River north to Cape York in northern Queensland, from the coast to about 4,000ft. The plant favours moist areas.

Variety *leroyi* of *C. madidum* is identical to the type in most respects, differing only in the labellum's lateral lobes, which are prominently recurved forwards, giving the labellum a very boat-shaped appearance. The variety also flowers later than the type form, from November to December, whereas the type flowers from August to October. Var. *leroyi* occurs only on the coast between the Barron and the Bloomfield Rivers in Queensland, the type of the variety coming from Emmagen Creek, north of Cape Tribulation in northern Queensland.

C. suave (Plate 93) differs greatly in appearance

Plate 91

Cymbidium canaliculatum   QUEENSLAND BLACK ORCHID

This distinctive colour form of tropical Queensland, long known as *v. sparkesii*, is no longer accorded varietial status. *C. canaliculatum* (**Banana Orchid**) extends from northern New South Wales through Queensland and Northern Territory into north-west Australia.

Plate 92

Cymbidium madidum   NORTHERN CYMBIDIUM

Throughout the high rainfall areas of tropical Australia, usually August flowering in the tropics, later in its southern occurrences.

from the other two species, having no or almost no swollen base or pseudo-bulb to the stem. But with age it forms a long elongate stem, the basal section covered with the remains of the old leaves. The dark green leaves are grasslike, much narrower and softer than in the other two species. The flower raceme is semi- to fully pendulous with numerous closely-packed flowers which tend to be slightly cupped, and are apple-green, light green, golden-green or brownish in colour. *C. suave* occurs from southern New South Wales north to northern Queensland and the Northern Territory, and is usually found on hardwood trees in open forest, also on *melaleuca* trees along water courses and at times on the margins of swamps, from sea level to at least 3,000ft. This species is sweetly scented and can be detected from a considerable distance in the bush.

The method this species used for propogation is by long adventitious growths which are produced at each knothole and crack in the tree.

Cymbidium is one genus of our Australian orchids that can readily be propagated commercially from seed. The home gardener can also raise a few seedlings in the following way: Fill a 3-in pot with sphagnum moss, cover this with a boiled piece of white towel, pushed down around the inside of the pot. Scatter the seed lightly on the towel. The pot is then watered from the bottom by standing it in a saucer of boiled water. The whole pot is then placed under a bell jar or bottle with the bottom cut off. Plants appear as little green blobs.

THE SUBTRIBE SARCANTHINAE

FOR MANY YEARS this section of the Australian orchids has caused the taxonomist and the general grower many identification difficulties, for it was obvious many did not fit the then known genera, and many plants assigned to different genera did not rightly belong there. Recently A. W. Dockrill published a review of this section which has placed some order in the ranks of Sarcanthinae.

In Australia Mr Dockrill classified twenty-one genera containing forty-four species. He deleted from Australia the genus *Sarcanthus* and most of the *Saccolabium* and has added a further four new genera. This treatment now clears up many of the anomalies and gives a much clearer structure for both expert and layman to work on.

Many of the Sarcanthinae respond well to cultivation and are generally grown on tree ferns or rolls of paperbark. They require a constantly moist atmosphere, as most are rain-forest species. The use of foliage nutriment sprays is most beneficial and can be used weekly in a diluted strength.

Judging by the arrangement of the column in the majority of the genera, insects are the chief pollinating agents.

Artificial hybridisation has been carried out mainly with one of the Australian genera, *Sarcochilus*. Some very interesting crosses have been made, including one between *Sarcochilus* and *Rhinerrhiza*, called by the raiser *Sarcorhiza*, but not validly described. Overseas hybridisers have given much attention to the exotic *Vanda* and *Phalaenopsis*, and have produced some truly beautiful orchids.

In the study of this subtribe one must concentrate on the intricate differences of the column and labellum structures, which on the whole are reasonably constant, and in general can be clearly seen

Plate 93

Cymbidium suave   SNAKE FLOWER

Extending from southern New South Wales to northern
Queensland also Northern Territory, the plant grows on
hardwood trees in open forest. August–October.

Plate 94

Sarcochilus australis   BUTTERFLY ORCHID

Ranging from Tasmania (where it was originally
discovered) through eastern Victoria to northern New
South Wales. October–November.

with the dissection of the labellum pouch and the use of an x10 hand lens. Some genera are immediately recognisable and can be easily assigned to the right genus, *e.g. Vanda, Phalaenopsis* while others, *e.g. Plectorrhiza erecta* and *Robiquetia tierneyana*, both similar in plant growth, have differences that are found mainly inside the labellum pouch. The former has a hairy callus in the front wall of the pouch, while the latter has no callus inside the pouch at all. It is features such as these, and the question whether the labellum has a column foot or a hinge directly on to the column that has provided difficulties in the past.

For this book it is proposed to divide the genera into three sections: (1) Plants without leaves. (2) Plants with leaves in a small tuft or group. (3) Plants with leaves on an extended stem.

In these divisions there are of course anomalies, but these will be noted in the correct place.

(1) This section contains only two genera, both small in habit and usually found on the limbs or boles of the trees.

Chiloschista phyllorhiza occurs in far northern Queensland and the Northern Territory, flowering in November to December. It shows a preference for trees close to salt water. It is a leafless plant consisting of numerous ribbon-like roots, which are flattened and rough, containing green chlorophyll. They apparently act like a normal leaf. The white, fragrant and short-lived flowers rise from the root axis in a slender raceme. The labellum is three-lobed with a pouch, and is hinged on a column foot. The pouch contains a finger-like hairy callus on the rear wall. An antenna-like appendage occurs on each side, at the top of the column.

Of the 170 species in the genus *Taeniophyllum*, only four are in Australia. The genus occurs from Ceylon to northern India, through Malaysia to Japan and from Tahiti to Australia. It is usually difficult to find a mature flowering plant in this genus, as they measure only 12cm over the full spread of the roots. The Australian species are *T. wilkianum*, *T. muelleri*, *T. lobatum* and *T. flavum*.

(2) This section contains ten genera with twenty-six species. By far the most conspicuous is *Phalaenopsis amabilis* var. *rosenstromii*, a beautiful Moth Orchid, with long racemes of white flowers. It occurs very sparingly on the Seaview Range and the Mossman and Daintree Rivers. It flowers from November to February. To grow *Phalaenopsis* requires a constant moist atmosphere, *i.e.* a heated glasshouse in the south. The Orange Blossom Orchid, *Sarcochilus falcatus* (Plate 95), is the type of the much-cultivated genus, and contains eleven species in Australia. *S. falcatus* is a shade-loving plant growing usually in moist gullies on softwood trees. The flowers vary in colour from white marked with yellow to orange, or a dusky rose. Plants have thick and longish leaves, the stems shortly scrambling.

S. fitzgeraldii and *S. hartmanii* have white flowers in an upright raceme, the segments marked reddish at the base. *S. fitzgeraldii* (Plate 96) is found in dark ravines in moss-beds and has a largish labellum with erect, falcate lateral lobes. *S. hartmanii* is found on cliff-faces and in leaf-mould in much more sun and has a small labellum for the size of the flower, with broad ovate-shaped lateral lobes. Fairy Bells, *S. ceciliae*, with its tuft of greenish spotted leaves and pink flowers, is hard to mistake. Also similar to *S. ceciliae* are *S. hillii* and *S. tricalliatus*. The sweetly-scented, green-flowered *S. olivaceus* is normally found on rocks. Its variety *borealis* from northern Queensland differs in its higher mid-lobe to the labellum.

S. dilatatus, also a northern New South Wales

Plate 95

Sarcochilus falcatus　　　ORANGE-BLOSSOM ORCHID

Extremely rare in eastern Victoria but extending through New South Wales to northern Queensland. June flowering in the north, October–November in the south.

Plate 96

Sarcochilus fitzgeraldii　　　RAVINE ORCHID

Confined from north-eastern New South Wales to southern Queensland, this orchid flowers in the late spring.

and Queensland plant, is similar to *S. australis*, differing in the mid-lobe of the labellum, which is almost absent. *S. australis* (Plate 94), the southernmost form of the genus, which extends from Tasmania through Victoria to northern New South Wales. A dubious species is *S. harriganae*, which has green flowers but is considered very close to *Parasarcochilus spathulatus*. Another rare species is *S. moorei*, a fairly large plant with yellow spotted flowers with broad clawed segments and leaves 12cm long that hails from Cape York Peninsula. It is also present in New Guinea and the Solomon Islands. Three plants that were at one time in *Sarcochilus* are *Parasarcochilus spathulatus, P. weinthalii,* and the recently-described *P. hirticalcar*. All resemble the small *Sarcochilus* in habit, but have a hollow spur to the labellum and lack the large callus at the base of the labellum.

Similar to *Sarcochilus* and at one time included in this genus is *Rhinerrhiza divitiflora* (back cover), a large plant with rough raspy roots. The flowers are long and spidery, bright orange, spotted with red. The bud remains very small until a few days before flowering, when it progresses rapidly. All flowers open simultaneously and rarely last longer than a day. The genus *Pomatocalpa* has about thirty species occurring from Burma through Malaysia, Indonesia, New Guinea, and the Pacific Islands to Australia, where we have only one species, *P. macphersonii* (Plate 98), a plant resembling *Rhinerrhiza* in habit. But the raceme of flowers is shorter and the flowers are much smaller. A small plant resembling *Parasarcochilus spathulatus* in habit is *Schistotylus purpuratus* from the Dorrigo Plateau area. Its narrow and curved leaves are about 4cm long. Up to eight pale green with purple-brown-marked flowers form a slender raceme. The white labellum has a hollow spur with a deflexed callus on the front wall at the top of the opening.

Similar to this species is *Papillilabium beckleri*, which differs mainly in that the hollow spur has no internal calli but a mere thickening of the front wall. *Saccolabiopsis armitii* from northern Queensland to Cape York is a small plant which prefers 'dry' scrub areas either on the coast or in the highlands. The raceme is dense with small green flowers. The spur of the labellum is hollow with no internal calli, while the mid-lobe extends forwards into another pocket or cup-shaped projection. The genus *Drymoanthus* has one species, in northern Queensland. A New Zealand species was called *Sarcochilus adversus*. The Australian *D. minutus*, a small plant, has a short raceme of flowers. The labellum is concave, fleshy and entire, with no spur but with a thickening at its apex only. No appendages are present in the cup-like labellum.

One of the three[47] *Robiquetia* species falls into this section; it is *R. rectifolia*, a small plant with a short stem like *Sarcochilus australis*, but with small green flowers with a few purple spots and a white labellum. The labellum in this genus has a hollow spur with no calli present while the mid-lobe recurves forwards prominently.

(3) Of the three other *Robiquetia* species in this section one, *R. wassellii* (Plate 99), has a short stem only 4cm long, but it can extend to 6cm. This species with the other two occurs in northern Queensland. *R. tierneyana* has a stem to 150cm. The leaves are thick and fleshy, and the flowers are the smallest of the three. Similar in plant habit to *R. tierneyana* is *Camarotis keffordii*, but the hollow labellum spur has an appendage low in the pouch on the front wall. The flower in this genus has the labellum uppermost. *Mobilabium hamatum* is a smaller plant with similar habit. The persistant leaves have a distinct hook at their apex. The flowering raceme is short with small flowers. The

Plate 97

Plectorrhiza tridentata   TANGLE ORCHID

The new generic designation of this orchid removes it from the genus *Thrixospermum* in which it was placed in 1958. Ranges from eastern Victoria to the Atherton Tablelands of Queensland. August–January.

Plate 98

Pomatocalpa macphersonii

Occurring in eastern tropical Queensland, this is the only Australian representative of this south-east Asian genus. Usually flowers July–October.

very mobile labellum has a solid spur with a raised callus between the lateral lobes. Closely resembling this genus is *Plectorrhiza*, a small genus of three species. The Tangle Roots, *P. tridentata* (Plate 97), has a twisted wiry stem with numerous tangled roots. It is a plant that rarely attaches itself firmly to anything, often hanging by one root over water-courses. The small highly-perfumed flowers have a spurred labellum at right angles to the column, and a finger-like callus attached to the front wall. *P. erecta* is similar in habit to *P. tridentata*, but has thick fleshy roots. *P. brevilabris* is a wiry pendulous plant with flowers longer and more acute than both other species; the labellum is in line with the column and longer than the former plant, with a hairy callus on the front wall. *Peristeranthus hillii* is a fairly robust plant endemic to Australia. The pendulous stems have thick broadly-rounded leaves that droop downwards from the stem, as does the densely flowered raceme, which is just longer than the leaves. The flowers are green with brown spots. The labellum is almost square in side-section; the solid interior has a conical callosite and the spur is short and broad. *Luisia teretifolia* (Plate 100) is confined to Cape York Peninsula and areas of the Northern Territory but extends to New Guinea and beyond. It has a preference for mangrove swamps and the greenish flowers with the dark red labellum are very seldom produced. The plant is sometimes branched and up to 40cm high; the leaves are terete (thin and round).

The genus *Thrixspermum* has two Australian species. *T. congestum* belongs to the section *Dendrocolla*, because the floral bracts are congested on a very short stem and face all directions. The plant is smallish and scrambling with few leaves, and the flowering raceme occurs in a terminal cluster. Flowers are smallish and the labellum (the interior of which is hairy) has a short sac at its base with a callus at the centre and another sac or pouch at its apex. *T. platystachys* is more robust, having the floral bracts fan-shaped. The flowers are larger than in *T. congestum;* the labellum sac is more distinct and the sac on the mid-lobe less pronounced. *Schoenorchis densiflora* is a slender scrambling plant with narrow leaves and a short spike of dense small flowers. The labellum is spurred at the base and the mid-lobe fleshy. The appendage inside the labellum is on the rear wall. The plant occurs in south-eastern Cape York Peninsula. *Vanda* in Australia contains two species, *V. tricolor* (Plate 85), a large strap leaf type with flowers to 7cm, occurring in the Northern Territory, Arnhem Land, and extending to Indonesia. It flowers in August to October. *V. whiteana* (front jacket) has smaller shining chocolate-flecked yellowish flowers to 3.5cm across, and occurs in Cape York Peninsula, flowering in November. A recent addition to this section is *Trichoglottis australiensis*. Its much branching habit is similar to *Plectorrhiza erecta*. The small clusters of flowers spring from stem-nodes. The labellum pouch has a tapered appendage at the mouth of the sac, attached to the rear wall. *Cleisostoma nugentii* is known only from the type description in *Queensland Flora*.

Plate 99

Robiquetia wassellii

This photograph is of the 'type' plant (i.e. the plant on which the original description was based) The original description was made in 1959.

Plate 100

Luisia teretifolia

Photograph of a Northern Territory plant previously recorded in Australia from coastal areas of Cape York, Queensland, also New Guinea and adjacent islands. November–February.

MISCELLANEOUS GENERA NOT PREVIOUSLY DISCUSSED

IN THE TEXT I have tried to describe as broad a cross-section as possible of the genera illustrated in the space available. The following brief notes give some data on those Australian genera not illustrated. Most of these plants are rare or very rare, the majority confined to northern Queensland or the tropics in general. Some have only been collected a few times during the past thirty or forty years. So far seventy of the eighty-five genera have been mentioned in some form, leaving fifteen genera with twenty-seven species to cover. Full details can be found in the recently published book *Australian Indigenous Orchids* by A. W. Dockrill, 1969.

Acriopsis javanica var. *nelsoniana*, north Queensland. Flowers July to October.

Anoectochilus yatesea, north Queensland. Flowers July to September.

Bromheadia venusta, Cape York Peninsula. Flowers June to August.

Cheirostylis ovata, northern New South Wales to Cape York Peninsula. Flowers June to September.

Corymborkis veratrifolia, north Queensland to New Guinea. Flowers December to March.

Goodyera viridiflora, north Queensland to Indonesia. Flowers June to August.

G. rubicunda, Atherton Tablelands to Cape York Peninsula to Malaysia. Flowers September to October.

Goadbyella, now obsolete.

Hetaeria oblongifolia, north Queensland to Philippine Islands. Flowers July to October.

H. polygonoides, north Queensland to New Guinea. Flowers June to August.

Malaxis latifolia, Queensland to India. Flowers December to April.

M. xanthochila, Cape York Peninsula. Flowers January to March.

Nervilia pachystomoides, north-east Queensland. Flowers unknown, only Holotype known.

N. discolor, north Queensland to Indonesia. Flowers December to February.

N. holochila, north-east Queensland across top of Australia to Western Australia. Flowers December to February.

N. aragoana, Arnhem Land, Northern Territory to India. Little-known. Flowers November to February.

N. uniflora, north-east Queensland, Arnhem Land and Northern Territory. Flowers November to February.

Pholidota pallida, north Queensland through to south China. Flowers March to May.

Phreatia crassiuscula, north Queensland. Flowers January or February.

P. baileyana, tropical Queensland. Flowers unknown.

P. robusta, north Queensland. Flowers October to February.

Spathoglottis paulinae, north-east Queensland to Cape York Peninsula. Flowers unknown.

Tainia parviflora, south-east Cape York Peninsula. Flowers September to November.

Pachystoma holtzei, north Queensland and Northern Territory. Flowers November to December.

Thelasis sp., unidentified plant now in Melbourne National Herbarium, Cape York Peninsula.

Zeuxine oblonga, north Queensland. Flowers July to September.

ACKNOWLEDGMENTS

AS THE text author, I would like to thank my co-author, Mr E. R. Rotherham, for much valued assistance, also my many corresponding friends, too numerous to mention individually, who have kindly supplied material for study.

Throughout this work I have drawn much data from a number of authors' works; these are credited below, with my thanks. Mr A. W. Dockrill; the late W. H. Nicholls; Mrs R. Ericson; Mr W. T. Upton; S. Clemesha and Dr R. Holtrum. For their co-operation I wish to thank: Director and staff members of the Herbarium, Royal Botanic Gardens, Sydney; Royal Botanic Gardens, Melbourne.

Finally I hope readers derive as much pleasure from the book as I have had compiling the text.

Leo I. Cady
Kiama, New South Wales

NOTES AND REFERENCES

1. Hatch, E. D. *Transactions of the Royal Society of New Zealand*, Parts 3 & 4 (1952) 400
2. Cady, L. I. *Australian Plants*, Vol. 5,2 (1969)
3. Sargent, O. *Annals of Botany*, Vol. XXIII
4. Coleman, Mrs E. *Victorian Naturalist*, March 1934
5. Blackmore, J. & Clemesha, S. *The Orchadian*, 2,12 (1968)
6. Forster, G. *Prodromus Florulae Insularum Australium* (1786) 59
7. Swartz, O. *Kongliga Svenska Velenskapsakademiens Handlinga Nya Foljd*, 21 (1800) 247
8. Cady, L. I. *Australian Plants*, Vol. 4,23 (1967) 163
9. Rogers, R. S. *Transactions of the Royal Society of South Australia*, Vol. 37 (1913) 51
10. Smith, J. J. *Exotic Botany*, 1 (1804)
11. Dockrill, A. W. *Victorian Naturalist*, 1981, September (1964) 128
12. Coleman, Mrs E. *Victorian Naturalist*, December (1932) February and May (1933)
13. Ericson, Mrs R. *Orchids of the West* (1951) 31
14. Forster, J. R. & G. *Characteres Generum Plantorum* (1776) 97, 8, T.49
15. Rogers, R. S. *Transactions of the Royal Society of South Australia*, Vol. 37 (1913) 56–60 Pl. 8–9
16. Rupp, H. M. R. *Orchids of New South Wales* (1943) 8
17. Hatch, E. D. *Transaction of the Royal Society of New Zealand*, 79 (1952) 396
18. Nicholls, W. H. *Orchids of Australia* (1951) Pl. 45
19. Nicholls, W. H. *Victorian Naturalist*, 66 (1949) 55–56
20. Coleman, Mrs E. *Victorian Naturalist*, April (1934)
21. Garnet, R. *Victorian Naturalist*, December (1940)
22. Rogers, R. S. *Transaction of the Royal Society of South Australia*, Vol. 33 (1909) 197
23. Dockrill, A. W. *Australian Orchid Review*, 21 March (1956) 26
24. Cady, L. I. *The Orchadian*, 2,2 (1965) 34
25. Rogers, R. S. *Victorian Naturalist* (1926) 179
26. Rotherham, E. *Victorian Naturalist*, 85 (1968) January
27. George, A. *Western Australian Naturalist*, 9,1 (1963) 3–9
28. Hooker, J. D. *Florae Nova Zealandiae* (1855)
29. Rogers, R. S. *Transactions of the Royal Society of South Australia*, Vol. LV (1931) 143
30. Sargent, O. *Journal of the Natural History Society of Western Australia*, 4 (1907) 6
31. FitzGerald, R. D. *Australian Orchids*
32. Nicholls, W. H. *Victorian Naturalist*, XLVII (1931) 157
33. Rupp, H. M. R. *Victorian Naturalist*, LVIII (1942) 188
34. Rupp, H. M. R. *Victorian Naturalist*, L (1933) 106
35. Summerhayes, V. *Wild Orchids of Britain* (1951) 315
36. Coleman, Mrs E. *Victorian Naturalist*, XLIII (131) 95
37. Cady, L. I. *Australian Plants*, 3,26 (1966) 245
38. Rogers, R. S. *Transactions of the Royal Society of South Australia*, Vol. 37 (1913) 48
39. Hatch, E. D. *Transactions of the Royal Society of New Zealand*, Vol. 82 Part 2 (1954) 613
40. Leeuwen, W. M. *Blumea Supplement*, 1,57–63 Fig. 3 Pl. 7 (1937)
41. Dockrill, A. W. *The Orchadian*, 1,10 (1963) 115
42. Nicholls, W. H. *Victorian Naturalist*, 67 (1950) 10
43. Dockrill, A. W. *Australian Plants*, 3 (1966) 328
44. For change of nomenclature see Rupp, H. M. R. *Orchids of New South Wales*, supplement (1969) 157
45. Plants of this section occur in Papua and recently A. W. Dockrill reports a plant of this section from northern Queensland.
46. Dockrill, A. W., *Australian Indigenous Orchids*, 1969
47. At date of publishing this number is five. For other two species see reference 46.

INDEX